# Biotechnology and Intellectual Property Rights

Kshitij Kumar Singh

# Biotechnology and Intellectual Property Rights

Legal and Social Implications

Kshitij Kumar Singh
Amity Law School
Noida
Uttar Pradesh
India

ISBN 978-81-322-2058-9   ISBN 978-81-322-2059-6 (eBook)
DOI 10.1007/978-81-322-2059-6
Springer New Delhi Heidelberg New York Dordrecht London

Library of Congress Control Number: 2014948601

© Springer India 2015
This work is subject to copyright. All rights are reserved by the Publisher, whether the whole or part of the material is concerned, specifically the rights of translation, reprinting, reuse of illustrations, recitation, broadcasting, reproduction on microfilms or in any other physical way, and transmission or information storage and retrieval, electronic adaptation, computer software, or by similar or dissimilar methodology now known or hereafter developed. Exempted from this legal reservation are brief excerpts in connection with reviews or scholarly analysis or material supplied specifically for the purpose of being entered and executed on a computer system, for exclusive use by the purchaser of the work. Duplication of this publication or parts thereof is permitted only under the provisions of the Copyright Law of the Publisher's location, in its current version, and permission for use must always be obtained from Springer. Permissions for use may be obtained through RightsLink at the Copyright Clearance Center. Violations are liable to prosecution under the respective Copyright Law.
The use of general descriptive names, registered names, trademarks, service marks, etc. in this publication does not imply, even in the absence of a specific statement, that such names are exempt from the relevant protective laws and regulations and therefore free for general use.
While the advice and information in this book are believed to be true and accurate at the date of publication, neither the authors nor the editors nor the publisher can accept any legal responsibility for any errors or omissions that may be made. The publisher makes no warranty, express or implied, with respect to the material contained herein.

Printed on acid-free paper

Springer is part of Springer Science+Business Media (www.springer.com)

# Author's Preface

Recent conjunction of biotechnology and intellectual property rights has long-term implications for law and society. Intellectual property laws that were framed in industrial age have proved to be insufficient in the current information age. In the present age, modern biotechnological inventions, particularly genetic inventions differ markedly from chemical and mechanical inventions that have been the traditional subject matter of patents. With the development of human genomics and success of Human Genome Project, gene becomes more important because of its informational content rather than its material qualities (physical attributes). Moreover, the emergence of bioinformatics and genomic databases has changed the face of biotechnology from lab-based technology to computer-based science, posing new challenges for intellectual property laws. In addition to legal implications, patents on gene and gene fragments have significant social and policy implications. Overbroad patent claims on genetic research tools and diagnostic genetic testing and aggressive licensing practices relating to them have serious implications for genetic innovation, health policies, patients' rights and society at large. In genetic research, increased extension of intellectual property rights to human genetic material may have an adverse impact upon the interests of research subjects from whom the human genetic material is extracted. Against this backdrop, the book analyses the legal and social implications arising from the conjunction of biotechnology and intellectual property rights, focussing particularly on human gene and genetic variations.

The book locates emerging legal, social and policy issues pertaining to biotechnology and intellectual property laws and suggests some meaningful solutions to them. The discussion in the book is streamlined to respond to few important questions: whether existing intellectual property laws at national and international levels can cope up with the challenges posed by biotechnology (especially genetic technology); whether aggressive assertion of intellectual property rights to genetic research tools, fundamental genetic research and human genetic resources stands in conflict with the rights of patients, independent researchers and research subjects; and whether open and collaborative biotechnology promotes genetic research and innovation. There are numerous books on intellectual property rights which deal with biotechnology, however, the present book provides a comprehensive overview of biotechnology and intellectual property rights and connects various aspects of the topic in an integrated manner, providing a fresh insight of law–biotechnology interface in tune with the current information age. It is aimed at providing basic and comprehensive knowledge pertaining to the topic to a wide range of audience comprising legal practitioners, law students, researchers and scholars interested in interdisciplinary research, policymakers and others interested in biotechnology and intellectual property rights.

The book is divided into seven chapters. Chapter 1 introduces the theme of the book and contains the background of the book, the concepts of biotechnology and intellectual property rights and the framework of the book. In Chap. 2, the book analyses the patent approaches of the USA, European Union, Canada and India on the basis of patent laws, administrative decisions and case law, bringing common points and differences among and between them. The book concludes that the selected countries for the study vary significantly in their approach to biotechnology in degree of patent protection and patent exclusions; however, all of them recognise patenting of biotechnology invention, given its commercial potential. In Chap. 3, the book analyses the international patent regime dealing with biotechnology, highlighting the potential gaps and uncertainties as to the scope of numerous terms such as invention, microorganisms, microbiological processes, essentially biological processes under TRIPS. It also discusses the impact of such uncertainties on developing countries given their relatively slow pace of scientific and technological development and the persistent conflict between developed and developing countries regarding the harmonisation of patent laws. Chapter 4 of the book undertakes the analysis of the social and policy implications of patents on genetic research tools and genetic testing and comes up with the conclusion that these concerns cannot be adequately addressed only by making changes in the patent systems as patent law is not expected to provide solutions to broad social and policy issues. It insists upon formulating policies and making legislations specific to genetic patents to regulate the patent practices such as patent licensing in order to provide viable solutions to such issues. The book analyses the ill effects of Myriad Genetics' patent claims on BRCA1 and BRCA2 gene, which prevents patients from taking a second opinion and verification testing. It concludes that in diagnostic field, exclusive licensing of genetic tests often obstructs the accessibility of genetic innovation or diagnostic genetic testing and advocates for non-exclusive licensing. In Chap. 5, the book examines the intricacies involved in providing effective intellectual property protection to bioinformatics and genomic databases and suggests a comprehensive review of existing intellectual property laws in the light of present information age. Keeping in view the collaborative nature of bioinformatics and genomic databases, the book evaluates the pros and cons of open biotechnology. The book analyses the extension of intellectual property rights to human genetic resources in the light of benefit sharing and informed consent in Chap. 6. It explains the ownership puzzle of human genetic material used in genetic research and suggests that ownership rights of research subjects in their extracted genetic material must be recognised. The book insists upon a careful application of intellectual property rights to human genetic resources. The concluding observations and possible way outs are provided in Chap. 7.

# Author's Preface

Despite the complex nature of the topic, the book approaches the issues pertaining to the topic in a clear, integrated and meaningful way. Though the analysis of the patentability of biotechnology in the book is limited to four jurisdictions, it gives fresh insights of biotech patent trends in different social, political and economic setups. It would be helpful in striking a balance between harmonisation and differentiation of patent laws. The analysis of social and policy implications of genetic patents is limited to available literature and supporting data. Since the science involved in biotechnology is of evolving nature, it is difficult to come up with definite solutions, however, the book provides an insight of law–biotechnology interface, highlighting emerging issues and providing some possible solutions to the existing problems.

In the process of writing this book, the support provided by the individuals and institutions is noteworthy. In this context, I most sincerely convey my deep sense of gratitude to my supervisor and guide in LL.M. and Ph.D., Prof. G. P. Verma, Law School, Banaras Hindu University (BHU), India, for his remarkable guidance and academic support during my work. As a supervisor, he has always encouraged me to produce quality work with his scholarly inputs. I am grateful to Prof. M. P. Singh, Law School, BHU, for igniting my thought process to cover some vital issues pertaining to my topic by his critical observations. I wish to express my sincere gratitude to Prof. Ali. Mehadi and Prof. R. K. Murali of Law School, BHU, for their encouragement and support they offered me during the work. I am grateful to Mr. Vinai Kumar Singh, Indian Society of International Law, New Delhi, for his great cooperation and support extended during the preparation of the book.

I take this opportunity to express my sincere regards to one of the most eminent scientists of India and great visionary, Dr. Lalji Singh, Vice Chancellor, BHU, who has always been a great source of inspiration for me regarding my academic pursuit. He encouraged me to work on law–technology interface. I am highly obliged to Prof. Mark Perry, who provided me an excellent environment to hone my research skills during my visit to Faculty of Law, University of Western Ontario, Canada, which helped immensely while writing my book. It was his guidance that broadened and advanced the level of my research and enabled me to develop global understanding of the subject. I am thankful to my friend Dr. Thomas Margoni, Institute of Information Law, Faculty of Law, University of Amsterdam, for sharing his thoughts over the topic of my book and encouraging my work. I am also thankful to my juniors Hemant, Bipin, Gaurav, Saurabh and Abhinav, who have been sincerely engaged with me during my work and extended their full support.

I sincerely acknowledge the support provided by the staff of Center for Cellular and Molecular Biology (CCMB), Hyderabad, India. I extend my thanks to the Library Staff, Law School, BHU, particularly Mr. Brijpal and Mr. Shobhnath for providing me valuable resources relating to my book. I must acknowledge the great support extended by the Library Staff of Faculty of Law, University of Western Ontario, Canada, Indian Law Institute, New Delhi, and Indian Society of International Law, New Delhi.

I am thankful to Sagarika Ghosh and Nupoor Singh at Springer India for their continuous support and cooperation and anonymous reviewers for Springer for their incisive and constructive comments.

Lastly, but most importantly, I acknowledge the contribution of my parents from the bottom of my heart, who have always pushed me to follow my dream and encouraged me to strive for academic excellence even in adverse situations. I wish to acknowledge and admire all kind of support and assistance provided by my siblings, Renu, Alok, Pooja and Prabhakar during my work.

NOIDA  
May 2014

Dr. Kshitij Kumar Singh  
Asst. Professor  
Amity Law School NOIDA  
Amity University U.P.

# Contents

| | | | |
|---|---|---|---|
| **1** | **Introduction** | | 1 |
| | 1.1 | Background | 1 |
| | 1.2 | Biotechnology and Intellectual Property Rights: A Conceptual Framework | 10 |
| | | 1.2.1 Biotechnology | 10 |
| | | 1.2.2 Intellectual Property Rights | 11 |
| | 1.3 | Nature, Purpose and Focus of the Book | 12 |
| | 1.4 | The Framework of the Book | 13 |
| | References | | 15 |
| **2** | **Patentability of Biotechnology: A Comparative Study with Regard to the USA, European Union, Canada and India** | | 17 |
| | 2.1 | Biotechnology and Patent Law | 17 |
| | | 2.1.1 Transformation of Biotechnology: From a Non-commercial Science to a Commercial Industry | 18 |
| | | 2.1.2 Conjunction of Biotechnology and Patent Law: Challenges Posed by Biotechnology Before the Existing Patent Systems | 19 |
| | | 2.1.3 Human Genetic Patents: A Special and Controversial Case of Biotechnology Patents | 20 |
| | | 2.1.4 Divergence in Biotechnology Patent Practices Among Different Jurisdictions | 22 |
| | 2.2 | Patentability of Biotechnology in the USA | 23 |
| | | 2.2.1 Biotechnology as a Patentable Subject Matter | 23 |
| | | 2.2.2 Other Statutory Requirements | 50 |
| | 2.3 | Patentability of Biotechnology in European Union | 64 |
| | | 2.3.1 Traces of a Unified System of Patents for European Union | 64 |
| | | 2.3.2 Specific Legislative Response to Biotechnology Inventions | 65 |
| | | 2.3.3 Sources Governing Patent Grants in Europe | 66 |
| | | 2.3.4 Biotechnology as a Patentable Subject Matter in European Union | 66 |
| | |     2.3.4.1 Patentable Subject Matter | 66 |
| | | 2.3.5 Other Statutory Criteria for Patents | 70 |

ix

|  |  |  |  | |
|---|---|---|---|---|
| | 2.4 | Patentability of Biotechnology Inventions in Canada | | 80 |
| | | 2.4.1 | Statutory Framework for Patenting | 81 |
| | | | 2.4.1.1 Patentable Subject Matter | 81 |
| | | | 2.4.1.2 Other Statutory Requirements for Patenting | 97 |
| | | 2.4.2 | Comparison of Canada with the USA and Europe | 98 |
| | 2.5 | Patentability of Biotechnology Inventions in India | | 99 |
| | | 2.5.1 | Dimminaco Case: Paving the Way for Biotechnology Patents in India | 100 |
| | | 2.5.2 | Statutory Provisions Regarding Biotechnological Inventions Under the Current Patent Act 1970 (as Amended in 1999, 2002 and 2005) | 102 |
| | | | 2.5.2.1 Biotechnological Inventions as Patentable Subject Matter | 102 |
| | | | 2.5.2.2 Other Statutory Requirements Under Indian Patent Act for Patenting | 104 |
| | | 2.5.3 | Status of Biotechnology Patent in India | 108 |
| | References | | | 108 |

## 3 Patentability of Biotechnology Under the International Patent Regime: Differentiation v. Harmonisation ... 111

| | | | | |
|---|---|---|---|---|
| | 3.1 | Territorial Nature of Patents | | 111 |
| | 3.2 | Internationalisation of Patent System: From Territorial to Global Patent Regime | | 112 |
| | 3.3 | Patentability of Biotechnology Under TRIPS: Interpreting TRIPS in the Light of Biotechnology Inventions | | 113 |
| | | 3.3.1 | Different Countries Interpret the Term 'Invention' Differently | 114 |
| | | 3.3.2 | Special Legislations for Different Technologies in Member Countries Violate Non-discrimination Provision Under TRIPS | 115 |
| | | 3.3.3 | Exceptions Under the TRIPS Agreement | 116 |
| | | 3.3.4 | Patenting of Life Forms Under the TRIPS Agreement: Internationalisation of Gene Patents | 118 |
| | | 3.3.5 | Article 27.3(b) of the TRIPS Agreement: A Temporary Compromise | 121 |
| | | | 3.3.5.1 GATT Negotiation | 123 |
| | | | 3.3.5.2 Review of Art. 27.3(b) of the TRIPS Agreement | 124 |
| | | 3.3.6 | Other Patent Eligibility Criteria Under TRIPS | 125 |
| | 3.4 | Feasibility of a Uniform Global Patent System: Differentiation v. Harmonisation | | 127 |
| | | 3.4.1 | Draft Substantive Patent Law Treaty | 129 |
| | | 3.4.2 | Differentiation vis-a-vis Harmonisation | 129 |
| | | 3.4.3 | Merits and Demerits of Uniform Patent Law | 130 |
| | | 3.4.4 | Relevance of the Existing International Patent Regime in the Present Technological Age | 131 |
| | | 3.4.5 | Tentative Harmonisation Efforts | 132 |

|     | 3.5   | Implications of Setting up a Uniform World Patent System | 134 |
|-----|-------|--|--|
|     | References | | 135 |
| **4** | **Legal, Social and Policy Implications of Genetic Patents: Issues of Accessibility, Quality of Research and Public Health** | | **137** |
|     | 4.1   | Commercialisation of Genetic Research and Its Impact on Academics | 137 |
|     | 4.2   | Importance of Patents in Genetic Research | 139 |
|     |       | 4.2.1 Impact of Patents with Broad Scope on Genetic Research | 140 |
|     |       | 4.2.2 Impact of Increasing Number of Gene Patents on Genetic Research: The Tragedy of Anticommons | 140 |
|     |       | 4.2.3 Patent Thickets | 141 |
|     |       | 4.2.4 Royalty Stacking | 141 |
|     | 4.3   | Patenting of Genetic Research Tools and its Impact on Research and Innovation | 142 |
|     |       | 4.3.1 Patentability of Genetic Research Tools | 143 |
|     |       | 4.3.2 Implications of Patents Relating to Genetic Research Tools for Society | 143 |
|     |       | 4.3.3 Patenting of ESTs and Reach Through Claims | 144 |
|     |       | 4.3.4 Impact of Patenting of Genetic Research Tools on Innovation | 144 |
|     | 4.4   | Common Practice Regarding Using Patented Research Tools in Public Sector and in Private Sector | 145 |
|     | 4.5   | Viable Options | 146 |
|     |       | 4.5.1 Exclusive Licensing Practices May Retard Innovation | 146 |
|     |       | 4.5.2 Non-exclusive Licensing over Genetic Research Tools Should be Encouraged | 147 |
|     |       | 4.5.3 Research Exemptions and Their Scope | 147 |
|     | 4.6   | Patenting of Genetic Tests for Diagnostic Purposes | 153 |
|     |       | 4.6.1 Myriad's Patents on BRCA1 and BRCA2 Genes: A Case Study | 154 |
|     |       | 4.6.2 Concerns Regarding Myriad's Patents on BRCA1 and BRCA2 Genes: Reactions Against Commercial Testing in the USA, European Union and Canada | 155 |
|     | 4.7   | Arguments in Favour of Patents on Diagnostic Tests—to Develop Diagnostic Tests Require Significant Efforts | 163 |
|     | 4.8   | Policy Implications of Myriad's Patents on BRCA1 and BRCA2 | 163 |
|     | 4.9   | The Possible Way Outs | 165 |
|     | References | | 167 |
| **5** | **Intellectual Property Protection to Bioinformatics and Genomic Databases and Open Source Analogy to Biotechnology** | | **169** |
|     | 5.1   | Transition in Biotechnology: From Lab-based Technology to Computer-based Science | 169 |
|     |       | 5.1.1 Definition of Bioinformatics | 170 |
|     |       | 5.1.2 Objection to the Extension | 170 |

| | 5.2 | Bioinformatics Databases | 171 |
|---|---|---|---|
| | | 5.2.1 Intellectual Property Protection to Bioinformatics | 171 |
| | |     5.2.1.1 Patentability of Bioinformatics Database | 171 |
| | |     5.2.1.2 Viability of Patent Protection with Respect to Bioinformatics Databases | 175 |
| | |     5.2.1.3 Copyright Protection to Bioinformatics Database | 176 |
| | |     5.2.1.4 EU Directive on Protection of Databases | 176 |
| | |     5.2.1.5 A Combination of Copyright Protection and Database Rights | 176 |
| | 5.3 | Bioinformatics Software | 177 |
| | | 5.3.1 Patent Protection to Software | 177 |
| | | 5.3.2 Copyright Protection to Software | 177 |
| | | 5.3.3 Trade Secret | 178 |
| | 5.4 | Intellectual Property Protection to Genomic Databases and Problem of Accessibility | 178 |
| | | 5.4.1 Goals of Genomic Databases | 179 |
| | | 5.4.2 Accessibility of Abstract Genomic Data: Current Standard | 179 |
| | |     5.4.2.1 Bermuda Principle | 180 |
| | |     5.4.2.2 Extension to Community Resource Projects | 181 |
| | |     5.4.2.3 Extension of Bermuda Principles to Phenotype Data | 182 |
| | |     5.4.2.4 Accessibility of Abstract (Post)Genomic Data: Developments | 182 |
| | |     5.4.2.5 The Limitations of a License | 183 |
| | 5.5 | Open Source Analogy to Biotechnology | 183 |
| | | 5.5.1 Nature and Scope of Open Biotechnology | 184 |
| | | 5.5.2 Difference Between Open Source Software and Open Source Bioinformatics Software | 186 |
| | | 5.5.3 Genomic Database Projects: The Human Genome Project as Open and Collaborative Genomic Database | 186 |
| | | 5.5.4 Importance of Open and Collaborative Databases | 187 |
| | | 5.5.5 Open Standards | 188 |
| | | 5.5.6 Free and Open Development | 189 |
| | | 5.5.7 Whether Open Development Is a Viable Business Model | 190 |
| | | 5.5.8 Is the Open Source Analogy Relevant to Biotechnology? | 191 |
| | | 5.5.9 Innovation and Open Development | 192 |
| | 5.6 | Is Open Bio Good for Developing Countries? | 192 |
| | References | | 193 |
| 6 | **Implications of Genetic Patents on Human Genetic Resources: Issues of Ownership, Benefit Sharing and Informed Consent** | | 195 |
| | 6.1 | Ownership of Human Genetic Material | 195 |
| | 6.2 | Ownership Rights of Research Subjects over Their Genetic Material Used in Genetic Research | 198 |

|  |  |  |  |
|---|---|---|---|
|  | 6.2.1 | Moore versus Regents of the University of California | 199 |
|  | 6.2.2 | Greenberg versus Miami Children's Hospital (Canavan Disease Case) | 203 |
|  | 6.2.3 | Washington University Versus Catalona | 206 |
|  | 6.2.4 | Havasupai Case | 210 |
| 6.3 | Ownership of Human Genetic Resources—from a Global Genetic Commons to National Property | | 214 |
|  | 6.3.1 | Genetic Resources: A Global Genetic Commons | 215 |
|  | 6.3.2 | Global Genetic Commons Tradition Eroded with the Expansion of Intellectual Property Rights over Genetic Material | 215 |
|  | 6.3.3 | The Reaction of Developing Countries Against the Extension of IPR over Genetic Material | 216 |
|  | 6.3.4 | Andean Common System on Access to Genetic Resources (Common System) | 217 |
|  |  | 6.3.4.1 Indian Position on Non-human Genetic Resource: The Biodiversity Act 2002 | 218 |
| 6.4 | Human Genetic Material | | 219 |
|  | 6.4.1 | International Law's Mishandling of Genetic Material and Its Implications | 219 |
|  | 6.4.2 | Sovereign Control versus Open System | 220 |
|  | 6.4.3 | Optimum Way Outs | 221 |
| 6.5 | Position of International Agreements on the Access of Human Genetic Resources and Benefit Sharing: Convention on Biological Diversity and TRIPS | | 222 |
|  | 6.5.1 | Benefit Sharing | 222 |
|  |  | 6.5.1.1 Andean Pact | 222 |
|  |  | 6.5.1.2 TRIPS | 223 |
|  |  | 6.5.1.3 United States National Bioethics Advisory Commission | 223 |
|  |  | 6.5.1.4 The International Bioethics Committee of UNESCO | 224 |
|  |  | 6.5.1.5 Convention on Biological Diversity | 224 |
|  |  | 6.5.1.6 Human Genome Organisation (HUGO) | 224 |
|  |  | 6.5.1.7 Indian Position on Benefit Sharing Under Biological Diversity Act | 225 |
| 6.6 | Possible Solutions | | 225 |
|  | 6.6.1 | Companies Incorporating Benefit-sharing Clauses in Their Strategies | 226 |
|  | 6.6.2 | Certificates of Origin for (Non-human) Genetic Resources | 226 |
|  | 6.6.3 | Whether an Amendment Required in Patent System or Outside of it | 226 |
| 6.7 | Need for Modification in the Existing Patent Regime | | 227 |
| References | | | 227 |

# 7 Conclusion and Suggestions ... 229
## 7.1 Suggestions ... 244
### 7.1.1 Chapter 2 ... 244
### 7.1.2 Chapter 3 ... 244
### 7.1.3 Chapter 4 ... 246
### 7.1.4 Chapter 5 ... 247
### 7.1.5 Chapter 6 ... 248
## References ... 249

# Index ... 251

# List of Abbreviations

| | |
|---|---|
| ACLU | The American Civil Liberties Union |
| AIA | American Invents Act |
| AIPLA Q.J. | American Intellectual Property Law Association Quarterly Journal |
| All ER (EC) | All England Law Reporter (European cases) |
| Art. | Article |
| ASU | Arizona State University |
| C. J. | Chief Justice |
| C.A. Fed. (Cal.) | Court of Appeals, Federal Circuit., California |
| C.A. Fed. (N.Y.) | Court of Appeals, Federal Circuit, New York |
| CAFC | Court of Appeal for Federal Circuit |
| CBAC | Canadian Biotechnology Advisory Committee |
| CBD | Convention of Biological Diversity |
| CDISC | Clinical Data Interchange Standards Consortium |
| CGIAR | Consultative Group on International Agricultural Research |
| CIHR | Canadian Institute of Health Research |
| CLIA | Clinical Laboratory Improvement Amendments |
| Co. | Company |
| Corp. | Corporation |
| CRP | Community Research Projects |
| Cust. & Pat. App. | Court of Customs and Patent Appeals Reports |
| D.C. | District of Columbia |
| DAS | Distributed Annotation System |
| DDBJ | DNA Database of Japan |
| DNA | Deoxyribonucleic Acid |
| DOE | Department of Energy |
| e.g. | Exampli Gratia |
| E.P.O.R | European Patent Office Reports |
| ed. | Edited by |
| edn. | Edition |
| eds. | Editors |
| EMBL | European Molecular Biology Laboratory |

| | |
|---|---|
| EMR | Exclusive Marketing Rights |
| en banc | In the bench |
| ENCODE | Encyclopedia of DNA elements |
| epo | Erythropoietin |
| EPO | European Patent Office |
| EST | Expressed sequence tag |
| et.al. | Among others |
| EU | European Union |
| F.T.R. | Federal Trial Reports |
| FAR | Federal Acquisition Regulations |
| FOSS | Free and Open Source Software |
| GATT | General Agreement on Tariffs and Trades |
| Hap Map | Haplotype Mapping Project |
| HC | High court |
| HGI | Human Genome Initiative |
| HGP | Human Genome Project |
| HIPAA | Health Insurance Portability and Accountability Act |
| HUGO | Human Genome Organisation |
| i.e. | That is |
| I.P.L.R. | Industrial Relations Law Reports |
| Ibid | Ibidem (in the reference immediately cited) |
| ICTSD | International Centre on Trade and Sustainable Development |
| Id | At the same |
| IDA | International Depository Authority |
| IMTECH | Institute of Microbial Technology |
| in re | In the matter of |
| Inc. | Incorporation |
| IP | Intellectual property |
| IPR | Intellectual Property Rights |
| IRB | Institutional Review Board |
| ITPGRFA | International Treaty on Plant Genetic resources for Food and Agriculture |
| JPO | Japan Patent Office |
| LDCs | Least developed countries |
| MCH | Miami Children's hospital |
| MPOP | Manual of Patent Office Practice |
| MPPP | Manual of Patent Office Practice and Procedure |
| MTAs | Material Transfer Agreements |
| MTCC | Microbial Type Culture Collection |
| N.D.Cal. | Northern District of California |
| NAFTA | North American Free Trade Agreement |
| NCHGR | National Center for Human Genome Research |
| NCIC | National Cancer Institute of Canada |
| NIH | National Institutes of Health |

| | |
|---|---|
| NISCAIR | National Institute of Science Communication and Information Resources |
| No. | Number |
| OBF | Open Bioinformatics Foundation |
| OECD | The Organisation for Economic Co-operation and Development |
| OHIP | Ontario Health Insurance Plan |
| OJEPO | Official Journal of the European Patent Office |
| OTA | Office of Technology Assessment |
| P.A.B | Patent Appeal Board |
| Para. | Paragraph |
| PBRA | Canadian Plant Breeders' Rights Act |
| dPCT | Patent Cooperation Treaty |
| PHOSITA | Person having ordinary skill in the art |
| PLT | Patent Law Treaty |
| pp | pages |
| PPF | Public Patent Foundation |
| PSA | Prostate-Specific Antigen |
| PTO | Patent and Trademark Office |
| Pvt. Ltd. | Private limited |
| PXE | Pseudoxanthoma elasticum |
| Quid pro quo | One thing in return for another |
| R&D | Research and development |
| RNA | Ribonucleic acid |
| S. Ct. | Supreme Court of United States |
| S.C.R. | Supreme Court Reports |
| S.D.N.Y. | Southern District of New York |
| SCC | Supreme Court cases |
| SDOs | Standards Development Organizations |
| Sec. | Section |
| SNP | Single nucleotide polymorphism |
| SPLT | Substantive Patent Law Treaty |
| Supra | Above or on an earlier page |
| TKDL | Traditional Knowledge Digital Library |
| TNF | The tumor necrosis factor |
| TRIPS | Trade-Related Aspects of Intellectual Property Rights |
| TUA | Technology Use Agreement |
| U.S | United States |
| U.S.C | United States Code |
| U.S.P.Q. | United States Patents Quarterly |
| UCLA L. REV | University of California, Los Angeles Law Review |
| UK | United Kingdom |
| UNCTAD | United Nations Conference on Trade and Development |
| UNESCO | Cultural Organization |
| UPOV | Convention on Protection of New Varieties of Plants |
| USPTO | United States Patent and Trademark Office |

| | |
|---|---|
| v. | Versus |
| Vol. | Volume |
| w.e.f. | With effective from |
| WIPO | World Intellectual Property Organization |
| WTO | World Trade Organization |

# About the Author

**Kshitij Kumar Singh** is an Assistant Professor at the Amity Law School, Noida, India. He obtained his Ph.D. (Law) from the Banaras Hindu University, Varanasi, India. During his doctoral study, he received the Canadian Commonwealth Scholarship Asia-Pacific 2010. The field of gene patenting and biotech law has been of special interest to him. Dr. Singh's LL.M. dissertation is on the topic "Human Genome and Cloning: Legal and Human Rights Issues"; he has also published many articles on biotechnology law. He gained related experience as a research intern (2009) and a visiting research fellow (2010) under the Canadian Commonwealth Scholarship 2010 at the University of Western Ontario, London, Canada. During this period he examined laws governing or needing biotechnology developments in India and Canada. The term was instrumental in expanding his research interest to the international aspect of the laws governing genetic patents and biotech research.

# Chapter 1
# Introduction

## 1.1 Background

Recent advances in biotechnology have brought dramatic transformations in the society. These transformations have been remarkable in the field of molecular biology and genetics, where biotechnology has opened new vistas for medicine and healthcare. The completion of the Human Genome Project has unravelled some mysteries of human life, reshaping our understanding of human genetics. With the increasing understanding of the genetic basis of human diseases through modern techniques such as diagnostic genetic testing, biotechnology has shown its potential to transform the basic framework of clinical medicine from one of 'diagnosis and treatment' to one of 'prediction and prevention'.[1] Modern techniques in biotechnology promise the new era of 'personalized medicines' with more accurate results as compared to traditional medicines. These scientific advances create commercial possibilities for industries. It started in 1970s with the introduction of new biological techniques such as the recombinant DNA technology, genetic engineering and cell culture which led to the emergence of the biotechnology industry. The enormous potential of biotechnological advances to produce commercial results has led to the conjunction of biotechnology and intellectual property rights (IPR), where industries and intellectual property developers seek to grab new biotechnological inventions through intellectual property protection.

The conjunction of biotechnology and IPR has posed unprecedented challenges before intellectual property laws. The existing intellectual property laws struggle to cope up with the challenges posed by biotechnological advances as they were framed in an age when these advances were not foreseen by the framers. The traditional doctrines of intellectual property laws have been extended to new subject matters such as genes, proteins and other unicellular and multi-cellular living organisms, which previously remained outside the grab of intellectual property law. Moreover, rapid advances in genomics and bioinformatics have raised the intellectual property protection debate for scientific information. IPR developers and holders claim that new technologies such as biotechnology fall within the existing

---

[1] Zimmern (1999).

bundle of the IPR, while end users assert that the technological change is so significant that contemporary intellectual property laws do not apply.

One of the most contentious issues in biotechnology-IPR discourse is the patentability of biotechnology. The traditional patent doctrines: patentable subject matter, novelty, non-obviousness (inventive step), utility (industrial applicability) and written description struggle while dealing with biotechnology inventions, especially genetic inventions. Human genes have become one of the most controversial subject matters of patent law because of its diverse nature. Since a gene comprises a number of elements, therefore, it is possible that a number of patents could be granted in relation to one gene. For instance, in relation to a particular gene, patents could be sought for the full sequence of a gene, an expressed sequence tag (EST), a single nucleotide polymorphism (SNP) or other variations of the gene. A gene has numerous uses as it can be a medically valuable product; an upstream research tool and vital information about the molecular basis of a disease.

One of the common objections against the gene patents is that genes are naturally occurring entities that are there to be discovered but not invented. In the context of gene patents, the line between discovery and invention is very thin and sometimes even discoveries are patentable through a broad interpretation of patent laws. With the development of human genomics and success of the Human Genome Project, a gene becomes more important because of its informational content rather than its material qualities. Here, the question arises, whether a gene as information is a patent eligible subject matter. Some critics see it as departure from the traditional patent doctrine, 'which is based on an agreement to disclose information in exchange of giving the inventor rights over material invention'.[2] They also argue that gene, being a patentable subject matter as information, would not only challenge the traditional patent system but also pose a great challenge to those who need access to information.[3]

The relevant novelty issue in gene patenting is whether the inventor has transformed the starting (pre-existing) natural materials into something new or simply removed imperfections from the original. Gene sequencing and gene isolation, once considered as laborious manual tasks have now become highly automated and routine parts of the laboratory exercise. This presents a great challenge to the inventive step/non-obviousness criterion as patent claims regarding these techniques seldom satisfy this criterion. In the case of gene patents, the utility criterion is also problematic as patents are being granted on gene fragments of unknown functions and gene sequences of limited or questionable utility. Since great uncertainty is involved in genetic technology, sometimes the description of an invention remains inconclusive. Sometimes patent claims are drafted in such a way that discloses only a correlation between a gene and a disease and the patent holders do not describe how the correlation was used to predict the disease.

The evolving jurisprudence of gene patents stems from the biotech patent practices of different countries. The scope and coverage of biotechnology patents vary

---

[2] World Health Organization (2005).
[3] *Ibid.*

from country to country. Even in countries having similar patent laws such as the USA and Canada, judicial interpretations by courts differ significantly. Judicial decisions rather than legislative efforts have shaped the fate of biotechnology patents in the USA and Canada. On the contrary, in Europe and India, significant legislative efforts have provided elaborated legislative provisions regarding biotechnology patents. Both the European Union and India have a list of patentable and nonpatentable subject matter in their respective legislations. Further, Europe and India contain the *ordre* public and morality clause to check the patentability of biotechnological inventions; however, the USA and Canada lack such provision in their patent laws. Patenting of a genetically engineered mouse called Harvard Oncomouse received different responses in the USA, Canada and Europe, which demonstrates diverse patent approaches towards higher life forms.

At the international level, the international patent regime struggles to cope up with the new challenges posed by biotechnology. It is due to certain ambiguities and potential gaps in the text of the Agreement on Trade-Related Aspects of Intellectual Property Rights (hereinafter TRIPS Agreement). The specific terms of the TRIPS Agreement set broad boundaries for the protection of biotechnological inventions. Article 27.1 of the TRIPS Agreement clearly states that patents should be granted for inventions in any field of technology without discrimination, subject to certain clauses. This provision gives legal mandate to biotechnology patents and eventually gene patents and obligates member states to accommodate biotechnological inventions. It is contended that the principle of equal treatment for all technologies as embodied in Art. 27.1 of the TRIPS Agreement undermines the unique nature of a gene and its distinct requirements. The term 'invention' under Art. 27.1 of the TRIPS Agreement is not defined and left open to the individual nation's interpretation. Further, Article 27.2 of the TRIPS Agreement allows member nations to exclude inventions from patentability for enforcing *ordre* public or morality; however, it does not stipulate the essence of such provision.

Article. 27.3(b) of the TRIPS Agreement specifically deals with biotechnological inventions, which excludes animals and plants from patentability but mandates patent protection for microorganisms and microbiological processes. However, the TRIPS Agreement neither defines the terms microorganisms and microbiological processes nor sets any parameter for determining the scope of these terms. This leaves great uncertainty as to the extent of patent protection for biotechnological inventions. Under the TRIPS, compelling signatory nations to provide patent protection to 'microorganisms' and 'microbiological processes', without defining the scope of the said terms, affects the interests of developing nations where the biotechnology field is not developed, when as signatory nations they are bound to provide patent protection to an all illusive term microorganism. Moreover, Art. 27.3(b) of the TRIPS Agreement provides that states can exclude from patentability 'essentially biological processes for the production of plants or animals'; however, it does not provide any parameter to ascertain which biological processes fall under the category of essentially biological processes.

The question also arises regarding the flexibility of the TRIPS Agreement. The TRIPS Agreement sets minimum standards before the member signatory nations to follow and generally requires that signatory countries provide broadly similar patent systems.[4] 'It does not delve much into the details of systems in part because the precise effects of those details are not known with certainty.'[5] To what extent the flexibilities under the TRIPS are helpful in dealing with biotechnological inventions remains a debatable issue.

Arguments are being made to harmonize patent laws of member nations by setting uniform standards beyond the TRIPS Agreement. The issue of harmonization of patent laws has generated much heated debates as establishing a single set of patent law is not only problematic but against the territorial nature of patents. Both uniformity and diversity has potentials and pitfalls. Uniformity simplifies the law, makes it easier to learn and describe, and reduces administrative costs. But, at the same time, uniformity brings certain disadvantages too as it makes the law unresponsive to local variations, eliminates inter-jurisdictional competition and decreases the possibilities for legal experimentation.[6] Diversity in patent laws is significant as it 'permits competition and breeds innovation'.[7] However, it does not check the free riding problem and cost burden. Therefore, 'the relevant policy question is to what extent inter-jurisdictional diversity and competition should be sacrificed to achieve global uniformity?'[8] There has been a political divide between developed and developing countries over this issue. Developed countries push for expanding the subject matter boundary by effectively eliminating exceptions. They advocate for the inclusion of a broader definition of patent eligible subject matter than the minimum set out under the TRIPS Agreement. Developing countries protest against this approach. The International Bureau of the World Intellectual Property Organization (WIPO) composed a draft Substantive Patent Law Treaty (SPLT) in 2001. However, disagreement between member states has put further discussion about the SPLT on indefinite hold. In the biotechnological context, creating a single set of patenting guidelines for the entire world has been proved to be very difficult given the controversy over issues such as patenting plants and animals.

In addition to legal implications, patents on gene and gene fragments have significant social and policy implications. These implications are related to the accessibility of genetic research tools, genetic innovation, health policies, patients' rights, clinical practice and the society at large. The potential of genetic research to produce commercial results prompted industries to collaborate with academic institutions in order to ensure commercial benefits through commercial agreements and patents. Governments have supported this trend because they believed that research carried by universities can contribute significantly in national economy. The trend of university-industry relationship began with the passage of the Bayh-Dole Act in the USA in 1980. The act gave universities the right to obtain IPR in inventions resulting from publicly funded research. It allowed industries to invest and col-

---

[4] Duffy (2002).
[5] *Ibid.*
[6] *Ibid.*
[7] *Ibid.*
[8] Smith (2000).

laborate in the university research and reap exclusive benefits through licensing and commercialization agreements. In other places such as Canada and Europe, where there was no law similar to the Bayh-Dole Act, universities committed themselves to enhance their commercialization outcomes.[9]

The commercialization of basic genetic research has threatened the free flow and open sharing of academic knowledge. Sceptics argue that patenting of upstream discoveries like DNA and genetic sequences would potentially block downstream research, which in turn blocks the development and access of new treatments. They maintain that the current commercialization process in the field of human genetics have made it more difficult for academic researchers to access and build upon scientific discoveries, and to openly disseminate their research results. The increased commercialization of upstream or basic genetic research has led to the patenting of gene fragments such as expressed sequence tags (ESTs) and single nucleotide polymers (SNPs), which are basically research tools. These research tools have no immediate therapeutic or diagnostic value but have a great value in conducting research. Patenting of these genetic research tools may stifle the genetic innovation.

In the field of human genetics, the breadth of patent claims and multiple numbers of patents on a single genetic innovation are the most problematic issues. Broad patents potentially limit opportunities for researchers to carry out further investigations on patented genes, to find out how they interact with other parts of the genome and any relationship they may have to particular diseases.[10] Multiple numbers of patents on a single gene create problems of anticommons, patent thickets and royalty stacking.

In the USA, Heller and Eisenberg described the problem of anticommons in the biomedical field as 'the tragedy of anticommons', where multiple patents on genes and gene fragments, which can be used as a tool for further research, discourage genetic innovation.[11] Access to these tools demands negotiation with all the patent holders, which can raise the transaction costs. This may result in underuse of resources. Patent thickets create a dense web of overlapping patents, where every patent holder has to find its way through the web in order to exercise his rights effectively. This again requires negotiation of multiple licenses, which may increase the transaction costs and retard the pace of innovation. A considerably large number of patents on genes and gene fragments may lead to royalty stacking problem. Royalty stacking arises when, in order to take a product to the market, the developer of the product takes licenses from all of the owners of the patents which affect the final product. These problems make the accessibility of research tools and platform genetic technologies more difficult for a researcher as he has to negotiate for a license from the patent holder.

Patenting of genetic testing, especially in the field of diagnostics has become a very controversial issue. Overbroad patent claims and aggressive licensing strategies impede the innovation process. Diagnostic gene patents also have serious im-

---

[9] Silverstein et al. (2009).
[10] Dutfield (2006).
[11] Heller and Eisenberg (1998).

plications for patients, researchers, physicians and the society at large. Patents in the diagnostic field inhibit the use of genetic diagnosis and the development of new drugs, giving rise to a conflict between patient's rights and patent rights.[12]

Myriad's patents on breast and ovarian cancer genes BRCA1 and BRCA2 reflect various social and policy implications involved in patenting of genetic testing. By using BRCA1 and BRCA2 genes, Myriad developed a method of diagnosing a predisposition for breast cancer and obtained patents in the USA, European Union, Japan and Canada. Myriad got the monopoly over the genetic diagnosis by maintaining arbitrary licensing policy, curtailing the access to the diagnostic technology. 'The U.S. BRCA patents are quite broad, covering a host of deleterious mutations in the BRCA1 and BRCA2 genes, the use of these mutations for diagnosis and prognosis for breast and ovarian cancer, screening for cancer pre-disposition, and the development of therapeutics to treat cancers with mutations in either gene.'[13] The company insists that all testing be performed by Myriad's own laboratories, charging exorbitant fee per test in spite of the fact that the same tests are available at a low cost. The company made it compulsory to send all DNA samples obtained from high-risk individuals for testing to its own laboratories in the USA. This allows Myriad to build up the only BRCA databank in the world, giving Myriad total control over the key research materials relating to genes coding for breast cancer susceptibility.

Myriad's BRCA tests have low predictive power for women without a family history of breast cancer. Moreover, the test offered by Myriad reportedly fails to detect 10–20 % of all expected mutations. 'The failure to detect such a large percentage of mutations in effect seriously jeopardizes the test quality and also significantly falls short of appropriate patient care when alternative more effective tests, could be readily available to the patient.'[14] Myriad does not allow any follow-up individualized genetic counselling regarding BRCA test results, preventing patients from obtaining a second opinion or alternative testing. This really undermines the appropriate quality of patient care. Further, since Myriad bars physicians and researchers to use the test without license, there remains little or almost no scope for improvement of the test. This impedes the training of the next generation of medical and laboratory geneticists, physicians and scientists in the area covered by the patent or license. Myriad patents have more deleterious impact on innovation as new scientific findings have shown that the one gene one disease concept is over and it is proved that one gene may be responsible for more than one disease; for instance, BRCA genes are no longer just associated with breast and ovarian cancer, but with prostate and pancreatic cancer too.[15]

Myriad faced strong opposition to the patenting of BRCA genes in Canada and Europe when it obtained the patent and began licensing of the patent to local

---

[12] *Ibid.*
[13] Bryn (2002).
[14] Paradise (2004).
[15] Allison (2009).

## 1.1 Background

companies.[16] Myriad's patents have been very controversial in Canada because of its impact on access to healthcare services and the potential cost of providing useful healthcare services within a publicly funded system. Recently, there has been strong reaction against Myriad's patent practice in the USA. In order to contend the social and policy implications of Myriad's gene patent practice, a lawsuit was filed on behalf of researchers, cancer survivors, breast cancer and women's health groups, and scientific associations representing geneticists, pathologists and laboratory professionals against the Myriad Genetics Inc. in the New York District Court. The plaintiffs were universities, professional organizations and cancer patients who could not afford to pay Myriad's monopoly price on diagnostics. The complaint charged that the patents-in-suit stifle diagnostic testing and research.

The US District Court for the Southern District of New York invalidated Myriad's patents covering mutations in BRCA1 and BRCA2, recognizing the growing concern about the impact of gene testing on the clinical practice and eventually on the rights of patients, researchers and other stake-holders.[17] The case reached to the Court of Appeal for Federal Circuit (CAFC) and eventually to the US Supreme Court, and it has been established by the Apex Court that 'naturally occurring DNA segment is a product of nature and not patent eligible merely because it has been isolated but cDNA is patent eligible because it is not naturally occurring.'[18] The court upheld the verdict of the CAFC that finding a link between a gene and a disease is not an invention denying the long-term patent practice. The case is significant from a policy perspective as many women are at risk for breast and ovarian cancer and they see genetic testing as one of the few preventative measures available.

In recent years, there has been enormous growth in the amount of genomic information due to significant genomic advances including the success of the Human Genome Project. Sometimes, the actual utility of most information is unknown, but needs to be preserved for future use. This necessitates devising methods to manage all the information gleaned so far and to arrange and catalogue them in a manner that may facilitate their use. Here, a fairly new discipline, bioinformatics,[19] comes into play which aims at the interpretation, integration and analysis of biological information including genetic information. The technology in bioinformatics comprises mostly programs and software which would aid in the compilation and updating of extensive databases of information, databases themselves, and also includes software which aid in the retrieval, analysis and comparison of relevant data.[20]

---

[16] Jones, *supra note* 13 at 138.

[17] *Association for Molecular Pathology v. United States Patent and Trademark Office*, Civil Action No. 09-4515 RWS (S.D.N.Y. 2009).

[18] *Association of Molecular Pathology et al. v. Myriad Genetics, Inc. et al.* 569 U.S. 12-1398 (Slip opinion) available at http://www.supremecourt.gov/opinions/12pdf/12-398_1b7d.pdf (last visited on 14 January 2014).

[19] The National Centre for Biotechnology Information defines it as 'field of study in which biology, information technology and computer science merge together to form a single discipline.' available at http://www.ncbi.nlm.nih.gov/About/primer/bioinformatics.html (last visited on December 16, 2011).

[20] Gopalan (2009).

Since a great amount of money and intellectual effort is involved in the study of bioinformatics (especially in genetic databases), intellectual property protection for bioinformatics becomes pertinent.

However, intellectual property protection varies according to the technology used in the bioinformatics field such as biological databases, algorithms, complex software etc. While some of these technologies may fit into the existing framework of intellectual property law, others may fall outside the scope of current legal protections. Databases per se, existing simply as collections or arrangements of raw data, are generally not patentable. It can be protected through copyright and sui generis database rights, where the former protects the compilation of database and latter the content of the same. However, providing an effective protection to genomic databases remains a challenge for copyright as well as database rights as they have certain inherent limitations.

Bioinformatics and genomic databases require significant collaborative efforts, therefore, need a continuously open and collaborative process for data collection and analysis. In such a situation, strict proprietary protection to genomic databases may conflict with public access to genetic information. Moreover, in the case of software and other bioinformatics research tools, it can be argued that free and open access to such tools might lead to the development of more efficient and streamlined methods of use, and to the discovery of new technologies or applications. These important aspects of bioinformatics are in direct conflict with the proprietary protection given to it. To remedy this situation, the most viable solution is seen in the open source biotechnology, inspired by the success of open source movement in the field of information technology.

However, unlike in information technology, where software is largely protected through copyright, products of biotech are usually protected through the patent system. Here, it remains doubtful whether the patent system can be used in the same successful manner as copyrights are used in the IT field to ensure open access. Moreover, the high cost and legal uncertainty associated with the genetic patents makes the biotech developers to choose alternative options in the form of copyrights, sui generis database rights, contractual agreements and commercial secrecy. Since biotech industries and even public research grant agencies are making huge investment in the development of an invention, so sometimes they see open biotechnology as antagonistic to IPR. Here, the relevant issue is whether it is possible that open-source software could be a viable option for commercial entities seeking profit in the bioinformatics field. Further, what would be the licensing strategies to ensure the open access to genomic databases?

The ownership of human genetic material remains another contentious issue as variety of proprietary rights are claimed over human genetic material. Although the fact that human genome is shared by the entire human race has prompted the United Nations to proclaim that human DNA is the Heritage of Humanity,[21] potential uses

---

[21] Art. 1 of *UNESCO's Universal Declaration on Human Genome and Human Rights 1997* proclaims that "[t]he human genome underlies the fundamental unity of all members of the human family, as well as the recognition of their inherent dignity and diversity. In a symbolic sense, it is the heritage of humanity."

of human DNA in development of diagnostics and therapeutics have led to patenting of human DNA, giving rise to patent claims over human genetic material. Since human genetic material promises future possible use, therefore, DNA is collected from human populations and contained in national bio-banks, where individual nations claim it as a national property. Human genetic material is also collected during medical treatment and research studies, where individual patients and research subjects stand up for their personal property rights in their blood, genes and data. Academic researchers, on the other hand, claim the human genetic material and data they have collected in the course of their research as their academic property.

Among these different claims for the ownership of human genetic material, the claims of patients and research subjects over their genetic material in medical treatment and genetic research have generated much heated debates. In genetic research, researchers and sponsors of clinical research are well placed to exploit the research carried on human genetic material of research subjects. The legal question of what ownership rights patients and research subjects have in their biological materials and their medical data is itself exceedingly ambiguous. Researchers contend that conferring ownership rights to research subjects and patients over their excised cells would impede the research and innovation processes. They also add that such materials become valuable due to their contribution as a scientist or researcher. Here, the question arises: Do research subjects or patients who have given their genetic material have no contribution at all? Further, whether may researchers obtain IPR through observation, isolation and manipulation of the human genetic material, without recognizing the contribution of the research subject, who contributed the genetic material in the first place?

The discussion relating to the ownership of human genetic material and benefit sharing is not confined to an individual and to a particular nation but it has gained utmost importance at the international level with the rapid increase in bio-prospecting. The human genome holds greater potential for drug development than genes of plants and animals. The main purpose of bio-prospecting is to find genetic materials that can be used to develop pharmaceutical, and other useful products that could be patented and marketed for profit. For the said purpose, researchers from academic and governmental institutions and private companies are paying much heed to population genetics and disease mechanisms. Here, 'the concern is that developing countries as a supplier of genetic material may end up having to pay high prices to the products that eventually developed from their own genetic material'.[22] The increased exploitation of human genetic resources through IPR has great implications for society. It is argued that potential exploitation of the DNA on all cell lines of indigenous people through patenting may give rise to increasing opposition to some vital population genetic studies and other works, beneficial to many countries.[23]

TRIPS does not contain any explicit reference to genetic material, and the laws that restrict access to genetic material to obtain remuneration for the member nation such as Convention on Biological Diversity (CBD) exclude human genetic material

---

[22] World Health Organization (2002).

[23] *Id*, at 143.

from their ambit. This has led to a growing exploitation of human genetic resources for scientific or commercial purposes. Objections are raised against permitting companies to profit from patented genes or goods utilizing such genetic material without compensating the donors of the underlying genetic material. Here, if the genetic material comes from a developing country, concerns regarding equitable sharing of benefits between developed and developing countries arise as well. Due to increased exploitation of human genetic resources of developing countries by developed countries, a trend towards viewing human genetic material as a natural or national resource is apparent.

## 1.2 Biotechnology and Intellectual Property Rights: A Conceptual Framework

### *1.2.1 Biotechnology*

Biotechnology is an evolving concept as its edifice is continuously expanding with the advancement of biotechnological advances. Over the period, the definition of biotechnology acquired a confusing status due to variety of interpretations. The Organization for Economic Cooperation and Development (OECD) has developed both a single definition of biotechnology and a list-based definition of different types of biotechnology. The single definition defines biotechnology as 'the application of science and technology to living organisms, as well as parts, products and models thereof to alter living or non living materials for the production of knowledge goods and services'.[24] The list-based definition contains a range of biotechnology techniques related to DNA/RNA, proteins and other molecules, cell and tissue culturing, process biotechnology techniques, gene and RNA receptors, bioinformatics and nano-biotechnology.[25] Scholars suggest that the single definition should always be accompanied by the list-based definition.

One important international definition is contained in the multinational environmental agreement, the CBD, which defines 'biotechnology' as: 'any technological application that uses biological systems, living organisms, or derivatives thereof, to make or modify products or processes for specific use'.[26] The Biosafety Protocol (The Cartegena Protocol on Bio safety to the CBD) defines 'modern biotechnology' as the application of:

> in vitro nucleic acid techniques, including recombinant deoxyribonucleic acid (DNA) and direct injection of nucleic acid into cells or organelles, or fusion of cells beyond the taxonomic family, that overcome natural physiological reproductive or recombination barriers and that are not techniques used in traditional breeding and selection.[27]

---

[24] Statistical Definition of Biotechnology (updated in 2005).
[25] *Ibid.*
[26] Art. 2 of the Convention on Biological Diversity, 1992.
[27] Art. 3(i) of the Cartegena Protocol on Biosafety, 2000.

In essence 'biotechnology is the application of biological system (animals, plants and microbial) to industries, agriculture and environment to produce better goods and services and enhance the wellbeing of mankind.'[28] It is the synergistic union of the biological sciences and technology based industrial art; the utilisation of the biological processes for the exploitation and manipulation of living organisms and biological systems in the development or the manufacture of a product or in the technological solution to a real world problem.[29] Finally, it can be concluded that 'biotechnology' is a term for a group of technologies that use biological matter or processes to generate new and useful products and processes.[30] Many suggest that the term should be used in a much broader sense to describe the whole range of methods, both ancient and modern, used to manipulate organic to reach the demands of human.[31]

## 1.2.2 Intellectual Property Rights

The term IPR is also a dynamic term with a continuously expanding scope. To put it simple, 'intellectual property rights are first of all property rights; secondly, they are property rights in something intangible and finally, they protect innovations and creations and reward innovative and creative activity.'[32] Intangible things or products which are derived from human intellectual activity constitute the subject matter of IPR. The human intellectual activities that commonly result in most intellectual properties are innovation and creativity.[33] Since exclusivity or exclusionary right is the hallmark of property, intellectual property is seen as an intangible subject matter emanating from the human intellect in respect of which a legal right of exclusivity may be granted.[34] This exclusionary right is termed as IPR, which is an artificial creation of law to secure exclusive possession of intellectual property. There are two underlying policies of intellectual property law: first to secure for the public the benefits of intellectual property, and second to regulate and manage competition.

It is always difficult to give a precise and uniform definition of IPR as the subject matter of these rights is continuously expanding. 'The definitional dimensions of intellectual property are further complicated by the fact that intellectual property regimes are the products of different philosophical and legal traditions.'[35] Intellectual property is used as a generic term, which came into regular use in the twentieth

---

[28] Noorzad (2000).

[29] Chawala (2005).

[30] "Key Issues in Biotechnology", United Nations Conference on Trade and Development 2002, available at www.unctad.org/en/docs/poitetebd10.en.pdf. (last visited on May 12, 2011).

[31] Bud (1991).

[32] Holyoak and Torremans (2005).

[33] Derek Bosworth & Elizabath Webster, *The Management of Intellectual Property* 25 (Edward Elgar Publishing Inc., Massachusetts, 2006).

[34] Holyoak and Torremans, *supra* note 32 at 27.

[35] Drahos (1999).

century. Initially there were two distinct terms in use: 'industrial property' and 'intellectual property'; the former was used to refer technology-based subject areas like patents, designs and trademarks while the latter used to refer copyright.[36] But, in the due course of time, this distinction has diminished and the modern convention is to use 'intellectual property' to refer to both industrial and intellectual property.[37]

Most definitions, in fact, simply list examples of IPR or the subject matter of those rights (often in inclusive form) rather than attempt to identify the essential attributes of intellectual property. An example of this approach is found in Article 2(vii) of the Convention Establishing the WIPO, signed at Stockholm on 14 July 1967.

The Convention Establishing the WIPO concluded in Stockholm on 14 July 1967 (Article 2(viii)) provides that 'intellectual property shall include rights relating to:

- literary, artistic and scientific works;
- performances of performing artists, phonograms and broadcasts;
- inventions in all fields of human endeavour;
- scientific discoveries;
- industrial designs;
- trademarks, service marks and commercial names and designations;
- protection against unfair competition;
- and all other rights resulting from intellectual activity in the industrial, scientific, literary or artistic fields.[38]

For the purposes of the TRIPS Agreement, 'intellectual property' refers to:

'... all categories of intellectual property that are the subject of Sections 1 through 7 of Part II of the agreement.'[39] This includes copyright and related rights, trademarks, geographical indications, industrial designs, patents, integrated circuit layout-designs and protection of undisclosed information.

Despite this categorisation, the canvas of IPR is expanding on an almost daily basis as new rights are created or existing rights are applied to fairly new subject matters such as genetic databases, semiconductor chips, human genes etc. Such expansion implicates the most basic tenets of society such as information, scientific data, entertainment and technology.

## 1.3 Nature, Purpose and Focus of the Book

In the line of the foregoing discussion, the book analyses the legal and social implications arising from the conjunction of biotechnology (specifically, genetic technology) and IPR. The study concentrates on a particular aspect of biotechnology i.e. the human gene. Since the traces of human gene patents are deeply rooted in the

---

[36] *Ibid.*
[37] *Ibid.*
[38] Art. 2(vii) of the Convention Establishing the World Intellectual Property Organization, 1967.
[39] Art. 1, para. 2 of TRIPS Agreement.

development of biotechnology patents as a whole, the present study carries discussion on biotechnology patents. For the purpose of the book, legal implications mean the challenges posed by biotechnology (especially genetic technology) before the existing IP laws. Social implications mean the wider implications of the genetic patents on the society, comprising various stake holders as patients, researchers, scientists, indigenous people and other social groups. The nature of the study is interdisciplinary, which focuses upon the interface of law and technology. In the discussion of the law-human genetic interface, ethical concerns are bound to come. These ethical concerns sometimes guide law to promote social good and reach legal excellence. Therefore, though the present study is primarily concerned with the legal and social implications, it also includes concerns relating to bioethics.

The topic of IPR is traditionally and to a large extent essentially concerned with questions inside a jurisdiction. Although the adoption and ratification of TRIPS has brought a unified character to intellectual property laws of member countries of World Trade Organization (WTO) to some extent yet these countries have adopted different approaches regarding gene patents, largely governed by their national policies. Keeping in view the divergence in intellectual property laws and their practice among various countries, the book analyses the intellectual property laws and trends relating to genetic technology of the USA, European Union, Canada and India. A detailed study of the USA is pertinent because being a pioneer in biotechnology research; it exerts great influence upon other countries. The European Union reflects the unified approach of different member states in a politically diversified system. The study of Canada becomes important because of its distinct approach regarding patenting of higher life forms despite having almost similar patent law to the USA. Canada reflects a nice blend of US and European approaches and is quite relevant in the context of India as a commonwealth country. Since India is in dearth of cases on gene patents and gene patent practices, it is always relevant to look into the case law and patent practices of other countries to develop a better understanding of the intricacies involved in the IPR debate relating to human gene. Since international agreements exert great influence on the intellectual property laws of member countries, the book also concentrates on international IPR regime dealing with genetic technology.

The book adds to the existing knowledge, giving fresh insights regarding the patent approaches of various countries to human gene patents. It analyses the potential gaps and ambiguities in international patent laws in the light of harmonisation and differentiation of patent laws. The book would be useful for India to develop better understanding of biotechnology patents by looking into the IP approaches of different countries and international practices and select the best, most appropriately suited to its own conditions.

## 1.4 The Framework of the Book

The book is divided into seven chapters. Chapter 1 introduces the topic in a lucid way, giving a proper legal and scientific background and connecting various aspects of the study. It contains the background of the book, conceptual framework of biotechnology and IPR, nature, purpose and focus of the book and the theme of the chapters.

Chapter 2 analyses the different patent approaches adopted by the USA, Canada, European Union and India regarding biotechnological inventions (especially genetic inventions) to bring about common issues and differences among these jurisdictions.

In Chap. 2, the author analyses the patent approaches of the USA, European Union, Canada and India on the basis of patent laws, administrative decisions and case laws, bringing common points and differences among and between them. The author comes up with the conclusion that the selected countries for the study vary significantly in their approach to biotechnology in the degree of patent protection and patent exclusions; however, the common point among them is that they all recognize patenting of biotechnology invention, given its economic value. He concludes that patent laws in the entire four jurisdictions struggle to cope up with new biotechnology inventions. In the light of such struggle, the author insists upon a comprehensive review of existing patent laws to address the genetic inventions in tune with the information age. He maintains that lack of such approach may prevent some useful inventions from society. As regards to the divergence in patent approaches of countries opted for the study, the author emphasises that the patent approach should always follow the socio-economic conditions of a particular country. He adds further that while making a distinction between patentable and non-patentable subject matter, the degree of human intervention should be considered.

In Chap. 3, the author analyses the international patent regime dealing with biotechnology, highlighting the potential gaps and uncertainties as to the scope of numerous terms such as invention, microorganisms, microbiological processes, essentially biological processes under TRIPS. He also discusses the impact of such uncertainties on developing countries given their relatively slow pace of scientific and technological development. The author explains the intricacies involved in providing an effective patent protection to new biotechnology inventions (that differ markedly from traditional subject matters of patents) at the international level in the light of technology neutral character of TRIPS.

Chapter 4 includes the study regarding the implications of patenting of genetic research tools and basic genetic research on the accessibility of genetic innovation. It discusses the viability of research exemption clauses under patent laws, and other relevant statutes regarding the accessibility of genetic research tools. The emphasis is on the patenting of genetic tests for diagnostic purposes and their impact on the rights of patients, researchers and other stakeholders. In this chapter, the author undertakes a detailed study of Myriad Genetics' patents on BRCA1 and BRCA2 genes, which prevents patients from taking a second opinion and verification testing. The author maintains that the social and policy implications of patents on genetic research tools and genetic testing cannot be adequately addressed only by making changes in the patent systems as patent law is not expected to provide solution to broad social and policy issues. He insists upon formulating policies and making legislations specific to genetic patents to regulate the patent practices such as patent licensing in order to provide viable solutions to such issues. The author adds that exclusivity provided by aggressive patent licensing strategies may not be in the public interest, and there is a continuing need for active defence of open science.

In Chap. 5, the author examines the intricacies involved in providing effective intellectual property protection to bioinformatics and genomic databases and suggests a comprehensive review of existing intellectual property laws in the light of the present information age. Keeping in view the collaborative nature of bioinformatics and genomic databases, the author evaluates the pros and cons of open biotechnology. He suggests that a variety of licensing schemes with or without intellectual property should be used to support the open nature of bioinformatics and genomic databases. The author adds that the intellectual property approach to bioinformatics should be balanced in such a way that it should not only incentivise the inventor or creator but also ensure the open and collaborative nature of bioinformatics.

In Chap. 6, the author analyses the extension of IPR to human genetic resources in the light of benefit sharing and informed consent. He explains the ownership puzzle of human genetic material used in genetic research and suggests that ownership rights of research subjects in their extracted genetic material must be recognized. Further, if researcher or sponsor conducting the research gains any benefit, the equitable sharing of that benefit must also be recognized. The author insists upon a careful application of IPR to human genetic resources. He also suggests that a clear distinction should be made between human genetic resources and non-human genetic resources and demands a specific legal approach to the former at the international level.

The concluding observations and possible submissions surfaced out of the study covered by the book are dealt in Chap. 7.

# References

Allison Malorye (2009) Diagnostic firms face new patent claim worries. Nature Biotechnology 27: 586

Andrew F. Christie (2006) A Legal Perspective. In: Bosworth Derek & Webster Elizabeth (2006) The Management of Intellectual Property. Edward Elgar Publishing Inc., Massachusetts

Bud Robert (1991) Biotechnology in twentieth century. Social Studies of Science 21: 444–445

Chawala H. S. (2005) Patenting of biological Material and Biotechnology. Journal of Intellectual Property Rights 10: 44

Duffy John F. (2002) Harmony and Diversity in Global Patent Law. Berkeley Technology Law Journal 7: 685–726 http://www.btlj.org/data/articles/17_02_02.pdf. Accesses 9 May 2012

Dutfield Graham (2006) DNA patenting: implications for public health research. Bulletin of World Health Organisation 84: 389

Gopalan Raghuvaran (2009) Bioinformatics: Scope of Intellectual Property Protection. Journal of Intellectual Property Rights 14: 48

Heller Michael A. and Eisenberg Rebecca S. (1998) Can Patents Deter Innovation? The Anticommons in Biomedical Research. Science 280: 698

Holyoak Jon & Torremans Paul (2005) Intellectual Property Rights. Oxford University Press, New York p. 11

Jones Bryn Williams (2002) History of Gene Patents: Tracing the Development and Application of Commercial BRCA Testing. Health Law Journal 10: 133

National Centre for Biotechnology Information (2011) Definition of Bioinformatics. http://www.ncbi.nlm.nih.gov/About/primer/bioinformatics.html. Accessed 16 December 2011

Noorzad Hazarat (2000) Biotechnology: its evolution, application, and environmental implications. www.mass.gov/envir/ota/pubs/biotech.pdf. Accessed 12 May 2011

Paradise Jordan (2004) European opposition to exclusive control over predictive breast cancer testing and the inherent implications for U.S. patent law and public policy: a case study of the Myriad genetics BRCA Patent Controversy. Food and drug law journal 59: 147

Peter Drahos (1999) The universality of intellectual property rights: origin and development http://www.wipo.int/edocs/mdocs/tk/en/wipo_unhchr_ip_pnl_98/wipo_unhchr_ip_pnl_98_1.pdf. Accessed 30 May 2011

Silverstein Tina, Joly Yann, Harmsen E. et al (2009) The commercialisation of genomic academic research: conflicting trend In: E. Richard Gold & Bartha Maria Knoppers (eds.) Biotechnology IP & Ethics LexisNexis Canada Inc., Markham, Ontario, p. 133

Smith Carrie P. (2000) Patenting Life: The potential and the pitfalls of using the WTO to globalize intellectual property rights. North Carolina Journal of International Law and Commercial Regulation 26: 143. www.international.westlaw.com. Accessed 18 May 2008

Statistical definition of biotechnology (2005). http://www.oecd.org/sti/biotech/statisticaldefinitionofbiotechnology.htm. Accessed 16 June 2011

United Nations Conference on Trade and Development (2002) Key issues in biotechnology. www.unctad.org/en/docs/poitetebd10.en.pdf. Accessed 12 May 2011

World Health Organization (2002) Genomics and World Health, Report of the Advisory Committee on Health Research 142 http://whqlibdoc.who.int/hq/2002/a74580.pdf. Accessed 15 June 2011

World Health Organization (2005) Genetics, genomics and the patenting of DNA-Review of potential implications for health in developing countries. http://www.who.int/genomics/FullReport.pdf. Accessed 20 May 2011

Zimmern R.L. (1999) The human genome project: a false dawn? British Medical Journal 319:1282

# Chapter 2
# Patentability of Biotechnology: A Comparative Study with Regard to the USA, European Union, Canada and India

Modern biotechnological advances have posed new challenges before the existing patent laws of countries as biotechnological inventions differ markedly from chemical and mechanical inventions that have been the traditional subject matter of patents. With the development of human genomics and success of the Human Genome Project, the gene becomes more important because of its informational content rather than its material qualities (physical attributes). Patent is a subject primarily concerned with questions inside a jurisdiction. Although the adoption and ratification of trade-related aspects of intellectual property rights (TRIPS) has brought a unified character to patent laws of member countries of the World Trade Organization (WTO) to a certain extent, these countries have adopted different approaches regarding biotechnology patents in tune with their national policies. As a result, the scope and coverage of biotechnology patents vary from country to country. Even in countries having similar patent laws such as the USA and Canada, interpretations of such laws by courts vary significantly. These variations among countries are important for the proper understanding of the trends in biotech patents. Therefore, the present chapter makes a comparative study of patent laws and practices relating to biotechnology patents in the USA, Canada, European Union and India in order to collate the common issues and the differences among and between them. The USA being a pioneer in biotechnology research exerts great influence upon other countries; the European Union reflects the unified approach of different member states in a politically diversified system; Canada makes a distinction between patenting of higher life forms and lower life forms and India represents the concerns of developing countries.

## 2.1 Biotechnology and Patent Law

Biotechnology and patent laws are not of recent origin; they have been present in our society for a long time. However, they became associated in recent years. This association became possible when biotechnology started creating commercial possibilities. Biotechnology, once primarily concerned with the academic field, has

© Springer India 2015
K. K. Singh, *Biotechnology and Intellectual Property Rights*,
DOI 10.1007/978-81-322-2059-6_2

been transformed into a commercial industry with an immense commercial potential. Recent bio-technological advances have presented unprecedented challenges before the existing patent laws, which have been slow to respond to technological challenges thus far.

### 2.1.1 Transformation of Biotechnology: From a Non-commercial Science to a Commercial Industry

Although the term 'biotechnology' gives an impression of modern, cutting-edge technology, its traces have remained present in early human settlements in the form of selective plant and animal breeding. Microorganisms have been employed for brewing and baking purposes for thousands of years. However, it was in early twentieth century, when the term biotechnology came into use. Karl Ereky, a Hungarian engineer, is said to have coined the term 'to refer to science and methods that permits products to be produced from raw materials with the aid of the living organisms'.[1] He first used the term *Biotechnologie* in a 1917 article (written in German) describing his pig fattening plant.[2] Taking the analogy of chemical technology, he suggested the word *Biotechnologie* to cover the area of technology associated with the living beings.[3]

Technology was generally associated more with chemistry and physics and less with biology, however, with the great advancements in biological sciences, this trend has anomalously changed.[4] Although from the 1880s analogies between physiological and technological structures did suggest a link between technological and biological evolution, however, biotechnology acquired a professional engineering dimension when the Americans took the biotechnics and biotechnology in the 1930s and 1940s.[5] Swedes emphasised on microbiology and Germans gave it institutional strength. America became a pioneer in the integration of molecular biology and engineering.[6]

The major breakthrough in the field of molecular biology and genetics was the discovery of deoxyribonucleic acid (DNA) in 1953 by Francis Crick and James Watson. It was considered as the discovery of secret of life. The discovery had been a grand success as the modern biotechnological advances in DNA technology have demonstrated that 'the DNA not only explains the very essence of every living cell but it promises great possibilities for future'.[7]

---

[1] Organisation for Economic Co-operation and Development (1999).
[2] Bud (1991).
[3] *Ibid.*
[4] *Id* at 444.
[5] *Id* at 444–45.
[6] *Ibid.*
[7] Yelpaala (2000).

Before the 1970s, biotechnology was primarily concerned with the academic development, intellectual curiosity, and expansion of scientific knowledge and the propagation of ideas for the benefit of humanity. Since the 1970s, with the advent of modern biotechnological techniques such as recombinant DNA technology, and tissue culture, researchers, venture capitalists and business community in general realised that DNA technology held the promise of significant financial rewards if the science could be converted into products or services. This realisation, in part, led to the emergence of the modern biotechnology industry of today.[8] Universities and research scientists once committed to the total openness were now interested in scientific discoveries that could be appropriated, protected within an intellectual property regime and eventually transformed into products and services in the market place.[9] This commercial trend led to a paradigm shift in biotechnological research from openness and sharing of knowledge and ideas to acquisitiveness and exclusivity.[10]

### 2.1.2 Conjunction of Biotechnology and Patent Law: Challenges Posed by Biotechnology Before the Existing Patent Systems

The Biotechnology industry is primarily made up of small, single product start-up companies. There is a close relationship between basic and applied science in the biotechnology field, and the biotechnology industry has a highly educated workforce.[11] Due to the close association between academic laboratories and industrial laboratories, biotechnology companies developed a culture that borrows several features of university setting.[12] The highly skilled work force required for the biotechnology industry can only be made available when the industry continues to attract academic scientists to the industry. Here, it becomes pertinent for the biotechnology industry to maintain a university like atmosphere and provide good economic incentives to the researchers, encouraging them to maintain a high level of innovation.[13] Further, due to the influence of academic research on biotechnology industry, the research ethos is encouraged with the encouragement of publication and sharing of results.[14]

Patents offer a viable option in this regard. Some form of economic incentive is *sine qua non* for the development of any start-up technological industry and patent offers such an incentive. Patents also encourage public disclosure of the invention so that society can be benefitted from that. Whether biotech patents fulfil the overall

---

[8] *Ibid.*
[9] *Ibid.*
[10] *Ibid.*
[11] Boyd (1997).
[12] *Ibid.*
[13] *Ibid.*
[14] *Ibid.*

social goal for which patents are intended is still a debatable issue but there is no doubt that patents are critical for the protection of biotechnology industry.[15] Due to the high level of uncertainty involved in biotechnology, investors are reluctant to invest into biotechnology ventures where the patent protection is lacking, or where the rights of patent holders are not clear.[16] This necessitates the conjunction of biotechnology and patent law.

Though patents are seen as an effective protection for biotechnology inventions, however, concerns have been made in recent years that current patent laws do not adequately encourage continued growth and research in biotechnology industry. Arguments have been made that a patent system designed to accommodate older technologies produces undesirable results when applied to a new and radically different technology such as biotechnology.

The commercial potential of biotechnology has led to the existing patent systems to accommodate fairly new subject matters such as DNA sequences, microorganisms, plants and animals which were not intended at the time of the framing of patent laws. Since a very high economic incentive has been involved with these subject matters, biotechnology industries and patent community have persuaded courts and legislatures that these subject matters should be treated no differently from mechanical and chemical invention.[17] In recent years, it has been realised that this analogy has failed to ensure a clear and adequate protection for modern biotechnology. Modern biotechnology differs significantly from chemical inventions with regard to structure and function and the manner and circumstances in which modern biotechnology and chemical inventions being created have been shown to differ markedly.[18] Apart from subject matter, modern biotechnological advances have posed new challenges before the existing patentability criteria such as novelty, non-obviousness, utility etc.

### 2.1.3 *Human Genetic Patents: A Special and Controversial Case of Biotechnology Patents*

There are groups which see the patenting of life forms such as human gene plainly wrong; there are some others who do not consider it necessarily wrong but in terms of its consequences. Sometimes the problem does not lie in the availability of patents but the way that granted patents are being asserted by the ruthless corporations on to the detriment of the public and especially vulnerable people like patients.[19] The opposition was driven by a variety of concerns including effects of such patenting on the environment, animal welfare, sustainable development, public health and

---

[15] Burk (1991).
[16] *Ibid.*
[17] Dutfield (2009).
[18] Pila (2003).
[19] Dutfield, *supra* note 17, at 192.

patient's rights. One of the most fundamental objections regarding gene patents is based on religious conviction—the notion that humans are 'Playing God'.[20]

As regards to patenting of gene, it is always contended that gene occurs naturally, hence is product of nature and not new. With rapid advancement in the field of molecular biology and genetics, gene sequencing once considered as a laborious manual task has become a highly automated and routine part of laboratory practice. This presents a great challenge to the inventive step/non-obviousness criterion. There is a significant challenge to the utility criterion as patents are being granted on gene fragments of unknown functions and gene sequences of limited or questionable utility. Since great uncertainty is involved in genetic technology, sometimes the description of an invention is not full. Many patents claimed far more than what the inventor actually discovered (e.g. claiming the sequence of a protein within the patent and then also asserting rights over all of the DNA sequences that encode that protein without describing those DNA sequences). The unique nature of science of genetics is the main reason for this failure.

Since a gene comprises a number of elements, therefore, it is possible that a number of patents could be granted in relation to one gene. For instance, in relation to a particular gene, patents could be sought for the full sequence of gene, an expressed sequence tag (EST), a single nucleotide polymorphism (SNP) or other variation of the gene, its promoter or enhancer, its individual exons or some other combination of the sequence.[21] Furthermore, a gene may be the subject of a product patent, process patent and use patents. For example, a product patent would cover the sequence itself which may be a product sold as a diagnostic tool to determine whether a particular gene is present. There could also be a product patent asserting rights over a gene and its product protein. The scope of the product patent is relatively wide as it asserts rights over all the uses of that product.[22]

A process patent may apply to some method of isolation and purification of a gene. As compared to product patents, process patent is unlikely to assert rights over the sequence of gene itself. However, if the gene or protein (which it encodes) is an element of a process or method that is used to produce some other product, the process patent may assert rights over the sequence of the gene.[23]

The use patent relates to a specific use of a gene. It could take the form of the use of a gene or part of its sequence in the manufacture of a medicine. It could also be framed in terms of the use of a gene for the diagnosis of a disease. The use patents in relation to gene and genetic components are very controversial due to their broad scope.[24] Commenting on the 'use patent' practice of Myriad over BRCA 1 the Nuffield Council of Bioethics observed:

> A broad use patent for a diagnostic test for BRCA1 that referred specifically to breast cancer would give the owner rights over all testing for that genetic susceptibility to breast

---

[20] *Ibid.*

[21] Cain (2003a).

[22] *Id.*, at 121.

[23] *Ibid.*

[24] *Ibid.*

cancer but not for other diseases. However, the effect of the patent owner having broad property rights over the diagnostic use of the gene for just one disease, would be that the patent owner has the monopoly over all ways of testing for that disease. This is because, even though the use patent does not include the sequence itself in the patent claims, in practice any other diagnostic test for the disease specified in a use patent would infringe that patent.[25]

So, the actual scope of gene patents depends upon the extent of the analysis carried out by the examiner at the relevant patent office. In addition to this with recent advancement in the field of genomics, gene has become more important as information rather than as a tangible entity. This transformation raises issue of patent eligibility of information, which has been excluded from patenting as 'scientific truths' and 'abstract ideas'.[26] Patenting gene as information has been viewed as departure from the long established patent practice.

### 2.1.4 Divergence in Biotechnology Patent Practices Among Different Jurisdictions

Although the minimum standards set out in international agreements brought some sort of uniformity in patent laws among member countries, however, the patent practices of these countries vary significantly. This is because the agreements provide considerable discretion to member countries in deciding how they choose to implement and operate their respective patent systems in tune with their respective needs.[27] The USA considers a much wider range of subject matter e.g. software, business methods and methods of medical treatment to be patentable as compared to Europe and Canada.[28] European patent law includes, for example, an 'ordre public and morality clause'[29] in its legislation that allows the European Patent Office (EPO) to exclude biotechnology patents for inventions, the commercialisation of which violates fundamental moral norms in Europe.[30] This clause has been interpreted to prevent the patenting of human embryonic stem cells by the EPO.[31] Indian patent law also contains a similar public order and morality clause.[32] However, Canadian and US law do not contain such clause, which provides them similar discretion to exclude patents on the basis of public order and morality.[33]

Patenting of whole animals and plants is another classic example of the discrepancies that exist between otherwise similar jurisdictions. These differences are

---

[25] Nuffield Council of Bioethics (2002).
[26] Merrill and Mazza (2006a).
[27] Gold (2009a).
[28] *Ibid.*
[29] Sec. 53(a) European Patent Convention, 1973.
[30] Gold and Knoppers, *supra* note 27, at 22.
[31] *Ibid.*
[32] Sec. 3(b), Patents Act, 1970.
[33] Gold and Knoppers, *supra* note 27, at 22.

explicit in the patenting of the Harvard College's genetically modified oncomouse, which received different treatments in the USA, Canada and Europe.[34] Further, three main patentability criteria—novelty, non-obviousness and utility—are applied more or less rigidly by different patent offices.[35] In order to understand the commonalities and differences in patent approaches of different countries regarding biotechnology inventions, a comparative study is pertinent. In this regard, the present study focuses on the patent approaches of the USA, European Union, Canada and India.

## 2.2 Patentability of Biotechnology in the USA

The USA has been a pioneer in the field of biotechnology and patent law. Initially, it has adopted relatively liberal approach while dealing with biotechnology patents but in due course of time it has developed its patent laws to fairly deal with the biotech challenges and the abuse of patent system in a matured way. It has therefore pioneered both the commercialisation of biotechnology applications and products and the development of patent law to protect them.[36]

The authority to grant patent is provided under the constitution of the USA. Congress is authorised 'to promote the Progress of Science and useful Arts, by securing for limited Times to Authors and Inventors the exclusive Right to their respective Writings and Discoveries'.[37] The basic requirement for obtaining a patent is set forth in Sections 101, 102, 103 and 112 of the Patent Act of 1952.

Section 101 prescribes the criterion for patentable subject matter as:

> Whoever invents or discovers any new and useful process, machine, manufacture, or composition of matter, or any new and useful improvement thereof, may obtain a patent therefore, subject to the conditions and requirements of this title.[38]

### 2.2.1 Biotechnology as a Patentable Subject Matter

*Traces of Biotechnology Patents in the USA Before 1980: Product of Nature Doctrine to Exclude Life Forms from Patenting* In the USA, patenting life forms was uncertain until 1980. Biotechnology products and processes were precluded from patenting and considered as product of nature. The product of nature doctrine implies that organisms or substances that occur in nature cannot be considered as

---

[34] *Ibid.*
[35] *Id.*, at 25.
[36] Dutfield, *supra* note 17, at 194–95.
[37] U.S. Constitution, Art. 1, Sec. 8, Cl. 8.
[38] 35 U.S.C. Sec . 101.

inventions and are therefore not patentable.[39] Very few patents were issued on 'mixtures or compounds that included microorganisms in modified form'.[40] It was only Pasteur's yeast culture product patent that exclusively covered living organisms.[41] In 1873, Louis Pasteur was granted a patent by the USPTO, claiming 'yeast free from organic germs of disease, as an article of manufacture'.[42] Nevertheless, since the 1880s USPTO apparently disallowed the patenting on any further life forms by applying product of nature doctrine.[43] The patent attorney, Grubb, rightly mentioned: 'In the USA, in spite of the precedent of the Pasteur patent…it has become practice of the Patent Office to refuse claims to living systems as not being patentable subject matter'.[44]

*The 1980s-Heralding a New Era of Wide Patenting of Biotechnology: Diamond versus Chakrabarty Case; Bayh-Dole Act 1980; Establishment of the Court of Appeal for Federal Circuit* In the year 1980 and onwards, there had been a sea change in granting patents over life forms, heralding a new era of biotechnology patents. Three major developments have contributed to this remarkable change: the US Supreme Court's decision in *Diamond versus Chakrabarty*[45] case; the Bayh-Dole Act; and the establishment of Court of Appeal for Federal Circuit (CAFC). The initial trend of USPTO to disallow patenting of life forms as a product of nature subsisted before the Supreme Court decision in *Diamond versus Chakrabarty*. Until the said decision, it was generally assumed within the emergent biotechnology sector that microorganisms could not be patented.[46] The Supreme Court decision in *Diamond versus Chakrabarty* allowed a patent on a new man made oil eating bacterium, paving way for biotechnology patents.[47]

The 1980 Bayh-Dole Act encouraged universities to patent, and thereby commercialise, inventions arising out of government sponsored research.[48] It has allowed public institutions to own inventions resulting from federally sponsored research and exclusive license to those inventions. Further, it requires from the institutions to establish patent policies for its employees, enabling them to seek patent protection of their invention.[49] This Act has provided a great deal of discretion to the institutions, encouraging the growth of biotechnology patents.

---

[39] Dutfield, *supra* note 17, at 195.

[40] *Id.,* at 195–96.

[41] *Id.,* at 196.

[42] Rimmer (2008).

[43] Dutfield, *supra* note 17, at 196.

[44] Philip Grubb, *Patents for Chemicals, Pharmaceuticals and Biotechnology*, 224–25 (Oxford: Oxford University Press, 4th edn., 2004), cited in Rimmer, *supra* note 42, at 24.

[45] *Diamond v. Chakrabarty, 447 U.S. 303 (1980).*

[46] Dutfield, *supra* note 17, at 196.

[47] *Id.,* at 195.

[48] Klein (2007).

[49] Boettiger and Bennet (2006).

The third major development was the establishment of the Court of Appeal for Federal Circuit (CAFC) by the Congress in 1982. The establishment was backed by a large group of high technology firms and trade associations in the telecommunications, computer and pharmaceutical industries, believing that a court devoted to patent cases would better represent its interest. The US government was also one of the main actors in the creation of CAFC as it was interested to support its science-based corporations by strengthening intellectual property protection worldwide.[50] Other underlying reasons were that the early Supreme Court decisions seemed to reflect an anti-patent mentality and lower courts had issued extremely inconsistent decisions on patent issues.[51] The CAFC has adopted a pro-patent approach which allows for the protection of biotechnology inventions. It has also issued decisions awarding high damages to patent owners in patent infringement cases, providing them additional protection.[52]

*Microorganisms as Patentable Subject Matter: Diamond versus Chakrabarty: a Case Study* Diamond versus Chakrabarty was the landmark case on life forms as patentable subject matter. The Supreme Court's decision in *Diamond versus Chakrabarty* has opened the patent field for biotechnology, giving boost to biotech industries. This outcome was a case of forum management achieved by a firm General Electric unrelated to biotechnology or pharmaceutical research.[53] By the time of filing the case the scientist involved, A. M. Chakrabarty, was in doubt that his genetically modified microorganisms could be patentable but the company lawyer, Leo MaLossi, was confident regarding the grant of patent. He was aware of the changing atmosphere that by now scientists understood living matter, including bacteria to be chemicals and Chakrabarty's bugs were patent eligible as new manufactures and compositions of matter.[54] Daniel J. Kevles describes the changing atmosphere as:

> By the time the case arrived at the court, it had become charged with the social and economic stakes that surrounded the swiftly accelerating commercialization of molecular biology. In the 1970s the new techniques of recombinant DNA were beginning to be exploited by adventurous start-ups such as Genentech. Companies were being founded at a rapid pace, while major pharmaceutical firms as well as several oil and chemical giants were plunging into recombinant DNA, initiating research programs of their own, letting research contracts to the start-ups, and even obtaining an equity interest in some of them.[55]

Though Chakrabarty had not used recombinant DNA technology to produce his oil eating bacterium, however, his case raised the vital issue of patentability of living organisms—an issue which was directly related to the fate of biotechnology patents and ultimately the future of biotech industry. This was reflected in ten amicus briefs

---

[50] Dutfield, *supra* note 17, at 200.
[51] Pila, *supra* note 18.
[52] *Ibid.*
[53] Dutfield, *supra note* 17, at 195.
[54] *Id.*, at 196.
[55] Kevles (2011)

filed by various economically interested organizations including Genentech, the Pharmaceutical Manufacturers Association, the American Patent Law Association, the New York Patent Law Association and the American Society for Microbiology.[56] Most of the amicus briefs supported General Electric's position.[57]

A. M. Chakrabarty filed a patent application in 1972 on behalf of General Electric, the abstract of the application titled, 'Microorganisms having multiple compatible degradative energy-generating plasmids and preparation thereof'.[58] There were 36 claims asserted in the application related to Chakrabarty's invention of 'a bacterium from the genus Pseudomonas containing therein at least two stable energy-generating plasmids, each of said plasmids providing a separate hydrocarbon degradative pathway'.[59]

The abstract of the application states:

> This human-made, genetically engineered bacterium is capable of breaking down multiple components of crude oil. Because of this property, which is possessed by no naturally occurring bacteria, Chakrabarty's invention is believed to have significant value for the treatment of oil spills.[60]

There were three types of patent claims; first the process claims for the method of producing bacteria; second, claims for an inoculum that comprised a carrier material floating on water, such as straw, and the new bacteria; and third, claims to the bacteria themselves.[61]

The patent examiner at USPTO allowed the first two groups but rejected the claims directed to the bacterium as unpatentable under 35 US Constitution (USC) Section 101. The USPTO rejected the claim on the bacterium on two grounds; first, bacteria were 'products of nature' and second, bacteria as 'living things' cannot be patentable.[62] Chakrabarty made an appeal against the rejection of these claims to the Patent Office Board of Appeals. The Board of Appeals reversed the first ground of rejection, mentioning that that the claimed bacteria were not products of nature, as they had been modified to produce a combination of plasmids that no known bacterium produced. It had, however, upheld the second ground that 'living things' could not be patented. While drawing the conclusion that Sec. 101 was not intended to cover living things such as these laboratory created microorganisms, the Board relied upon the legislative history of the 1930 Plant Patent Act, in which Congress extended patent protection to certain asexually reproduced plants.[63]

---

[56] *Ibid.*

[57] *Ibid.*

[58] A. Chakrabarty, 'Microorganisms having multiple compatible degradative energy-generating plasmids and preparation thereof', (1972) U S Patent No: 4,259,444.

[59] *Ibid.*

[60] 447 US 306 (*Chakrabarty*).

[61] *Id*, at 306–307.

[62] 447 US 307 (*Chakrabarty*).

[63] *Ibid.*

## 2.2 Patentability of Biotechnology in the USA

*Living and Non-living Distinction for Patentable Subject Matter* In 1978, an appeal was made to the Court of Customs and Patent Appeals. The court overturned the rejection made by Board of Appeals stating that 'the fact that microorganisms were alive was without legal significance for purposes of the patent law'.[64] Rich J. held that the claims were not outside the scope of patentable inventions merely because they were drawn to 'live organisms'.[65]

On reconsideration, the same court referred to its earlier decision in the *Application of Bergy*,[66] which involved patent claim relating to a biologically pure culture of the microorganism streptomyces vellosus.[67] Rich J. reversed the decision of rejection by the USPTO and held that it was in the public interest to include microorganisms within the terms of manufacture and composition of matter. Rich J. compared microorganisms with chemical elements, compounds and compositions of matter which are not considered to be alive despite their capacities to react and promote reaction to produce new compounds and compositions by chemical processes.[68] On remand, he upheld his previous stand that the claim cannot be rejected on the sole ground that it was for 'living organism'.[69]

The litigation reached to the US Supreme Court when Sidney Diamond, the Commissioner of the USPTO, sought and won a writ of certiorari from the said court.[70] Apart from the submissions from petitioners and respondents, the court received *amicus curiae* from different interested parties. The Supreme Court in this case held by a majority 5:4 decision that 'A live, human-made micro-organism is patentable subject matter under Sec. 101. Respondent's microorganism constitutes a "manufacture" or "composition of matter" within that statute'.[71]

In giving the wide interpretation of the terms 'manufacture' and 'composition of matter', the court looked into the prior decisions. *Perrin versus United States*[72] guided that in making statutory interpretation 'unless otherwise defined, words will be interpreted as taking their ordinary, contemporary, common meaning'.[73] *United States versus Dubilier Condenser Corp.*[74] indicated that courts 'should not read into

---

[64] *Application of Chakrabarty* 571 F.2d 40 Cust. & Pat. App. (1978) cited in Rimmer, *supra* note 42, at 28.

[65] Ibid.

[66] *Application of Bergy* 563 F.2d 1031 Cust. & Pat. App. (1977).

[67] M. Bergy, J. Coats and V. Malik, 'Process for preparing lincomycin', (1974) US Patent Application No: 477, 766 cited in Rimmer, *supra* note 42 at 29.

[68] *Application of Bergy* 563 F.2d 1031 at 1038 Cust. & Pat. App. (1977) cited in Rimmer, *supra* note 42, at 29.

[69] *Application of Bergy* 596 F.2d 952 Cust. & Pat. App. (1979) cited in Rimmer, *supra* note 42, at 29.

[70] Rimmer, *supra* note 42, at 30.

[71] 447 U.S. 308–318. (*Chakrabarty*).

[72] 444 U.S. 37 (1979).

[73] *Id.*, at 42.

[74] 289 U.S. 178 (1933).

the patent laws limitations and conditions which the legislature has not expressed'.[75] Based upon these judicial constructions Berger C. J. opined:

> Guided by these cannons of construction, this court has read the term, 'manufacture' in [Sec.] 101 in accordance with its dictionary definition to mean 'the production of articles for use from raw or prepared materials by giving to these materials new forms, qualities, properties, or combinations, whether by hand labour, or machinery'. Similarly, composition of matter' has been construed consistent with its common usage to include 'all compositions of two or more substances and… all composite articles, whether they be the results of chemical union, or of mechanical mixture, or whether they be gases, fluids powders or solids'.[76]

*Scope of the Patent Laws: Anything Under the Sun Made by Man* Regarding the scope of patent law, Berger C. J. held that 'In choosing such expansive terms as "manufacture" and "composition of matter", modified by the comprehensive "any", Congress plainly contemplated that the patent laws would be given wide scope'.[77] He added that the relevant legislative history also supports a broad construction and referred to the Patent Act of 1793, which embodied its author, Thomas Jefferson's philosophy that 'ingenuity should receive a liberal encouragement'.[78]

The Act defined statutory subject matter as 'any new and useful art, machine, manufacture, or composition of matter, or any new or useful improvement [thereof]'.[79] This broad language remained intact in subsequent patent statutes in 1836, 1870 and 1874. In 1952, when the patent laws were recodified, Congress replaced only the word 'art' with 'process', keeping the remaining Jefferson's language intact.[80] Berger C. J. had provided a much broader scope to the patent laws by referring to the Committee Reports accompanying the 1952 Act: 'The Committee Reports accompanying the 1952 Act inform us that Congress intended statutory subject matter to "include anything under the sun that is made by man"'.[81]

*Limitations on the Scope of Patentable Subject Matter* While interpreting the scope of the statutory subject matter broadly to include 'anything under the sun made by man', the Supreme Court did not rule out limitations to such scope. The court recognized certain exclusions such as the laws of nature, physical phenomena and abstract ideas, which have been held not patentable in several cases.[82] The court maintained:

---

[75] Id., at 199.

[76] 447 U. S. 308 (*Chakrabarty*).

[77] *Ibid*

[78] *Id.*, at 308–309, quoting 5 Writings of Thomas Jefferson 75–76 (Washington ed. 1871).

[79] Act of Feb. 21, 1793, Sec. 1, 1 Stat. 319.

[80] 447 U.S. 309 (*Chakrabarty*).

[81] *Ibid*, referring S Rep. No 1979, 82d Cong., 2d Sess., 5 (1952); H.R.Rep. No. 1979, 82d Cong., 2d Sess., 6 (1952).

[82] 447 U.S. 309 (*Chakrabarty*), referring Parker v. Flook,437 U.S. 584 (1978); Gottschalk v. Benson, *409 U. S. 63, 409 U. S. 67 (1972)*; Funk Brothers Seed Co. v. Kalo Inoculant Co., 333 U. S. 127, 333 U. S. 130 (1948); O'Reilly v. Morse,15 How. 62,56 U. S. 112-121 (1854); Le Roy v. Tatham,14 How. 156,55 U. S. 175 (1853) Le Roy v. Tatham,14 How. 156, 55 U.S. 175 (1853).

## 2.2 Patentability of Biotechnology in the USA

Thus, a new mineral discovered in the earth or a new plant found in the wild is not patentable subject matter. Likewise, Einstein could not patent his celebrated law that $E=mc2$; nor could Newton have patented the law of gravity. Such discoveries are "manifestations of... nature, free to all men and reserved exclusively to none".[83]

In the light of such exclusions, the court found Chakrabarty's microorganisms plainly patentable:

> Judged in this light, respondent's micro-organism plainly qualifies as patentable subject matter. His claim is not to a hitherto unknown natural phenomenon, but to a non-naturally occurring manufacture or composition of matter—a product of human ingenuity "having a distinctive name, character [and] use."[84]

In *Chakrabarty*, the Supreme Court did not concur with the then prevailing product of nature doctrine, propounded by it in the case of *Funk Bros Seed Co versus Kalo Inoculant Co*.[85] The court made a distinction between the Chakrabarty's claims for the genetically engineered microorganism to *Funk Brothers Seed Co. versus Kalo Inoculant Co.*,[86] in which the claimant had discovered certain naturally occurring bacteria, useful for agricultural purposes, could be combined into a single package without adverse effects and sought a patent on the packaged bacteria. The court in *Funk Brothers'* Case ruled the packaged bacteria non-patentable by concluding that the patentee had discovered 'only some of the handiwork of nature'.[87]

On the other hand in *Chakrabarty*, the Supreme Court upheld the patent on Chakrabaty's genetically engineered microorganisms because he had significantly manipulated nature. The court stated:

> Here, by contrast, the patentee has produced a new bacterium with markedly different characteristics from any found in nature, and one having the potential for significant utility. His discovery is not nature's handiwork, but his own; accordingly it is patentable subject matter under [Sec.] 101.[88]

The Supreme Court rejected the argument made by petitioner that the passage of sui generis legislations, the 1930 Plant Patent Act, which afforded patent protection to certain asexually reproduced plants and the 1970 Plant Variety Protection Act, which authorized protection for certain sexually reproduced plants but excluded bacteria from its protection, evidences congressional understanding that the terms 'manufacture' or 'composition of matter' do not include living things; if they did, neither Act would have been necessary.[89]

After looking into the legislative history of the said Acts, the court observed:

---

[83] 447 U.S. 309 (*Chakrabarty*), quoting *Funk Brothers Seed Co. v. Kalo Inoculant Co., 333 U. S. 130 (1948)*.

[84] 447 U. S. 309–10, quoting *Hartranft v. Wiegmann,* 121 U. S. 609, 121 U. S. 615 (1887).

[85] 333 U. S. 127, 333 U. S. 130 (1948).

[86] 333 U. S. 127, 333 U.S. 130 (1948).

[87] 333 U.S. 131 (1948).

[88] 447 U.S. 310 (*Chakrabarty*).

[89] *Id.,* at 311.

> In enacting the Plant Patent Act… Congress thus recognised that the relevant distinction was not between living and inanimate things but between products of nature, whether living or not and human made inventions. Here respondent's microorganism is the result of human ingenuity and research. Hence the passage of the Plant Patent Act affords the Government no support. Nor does the passage of the 1970 Plant Variety Protection Act support the Government's position. As the Government acknowledges, sexually reproduced plants were not included under the 1930 Act, because new varieties could not be reproduced true-to-type through seedlings. By 1970, however it was generally recognised that true-to-type reproduction was possible and that plant patent protection was therefore appropriate. The 1970 Act extended that protection. There is nothing in its language or history to suggest that it was enacted because Sec. 101 did not include living things.[90]

The petitioner argued that since genetic technology was unforeseen when Congress enacted Sec. 101, therefore the resolution of the patentability of inventions such as respondent's should be left to Congress.[91] Rejecting the petitioner's argument, the court maintained:

> The unambiguous language of [Sec.] 101 fairly embraces respondent's invention. Arguments against patentability under [Sec.] 101, based on potential hazards that may be generated by genetic research, should be addressed to the Congress and the Executive, not to the Judiciary.[92]

Expressing the dissenting opinion on behalf of White J., Marshall J. and Powell J., Brennan J. emphasised that the courts should be differential to the wishes of the Congress, and extend protection no further than statute provides.[93] He maintained that in absence of legislative direction, 'the Court should leave the Congress the decisions whether and how far to extend the patent privilege into areas where the common understanding has been that patents are not available'.[94]

Referring to the Plant Patent Act 1930 (US) and the Plant Variety Protection Act, 1970 (US), Brennan J. noted: 'In these two Acts Congress has addressed the general problem of patenting animate inventions and has chosen carefully limited language granting protection to some kinds of discoveries, but specifically excluding others'.[95] He concluded: 'These Acts strongly evidence a congressional limitation that excludes bacteria from patentability'.[96] Brennan J. also maintained: 'It is the role of Congress not this Court, to broaden or narrow the reach of the patent laws'.[97]

*Implications of Diamond versus Chakrabarty on the Scope of Biotechnology Patents* The Supreme Court decision in *Chakrabarty* received mixed reactions; while biotechnology industry groups and biotechnology advocates welcomed the decision, however, biotechnology patent opponents involving various non-governmental

---

[90] *Id.,* at 312.

[91] *Id.,* at 314.

[92] *Id.,* at 304, 314–18.

[93] Rimmer, *supra* note 42, at 42.

[94] 447 U.S. 319 (*Chakrabarty*).

[95] *Ibid.*

[96] *Ibid.*

[97] *Ibid.*

## 2.2 Patentability of Biotechnology in the USA

organisations and anti-biotech activists criticized the decision for extending patent protection to life forms. Cary Fowler, then an anti-biotech patenting activist, describes the situation:

> The GE- Chakrabarty case was a major tactical victory for the industry. Not only did it secure the protection it has long sought, but it did so through a new arena, the court system. The courts were neither fast nor cheap, but they were faster and cheaper than the political process. They were also a foreign territory to most advocacy groups.[98]

This decision of the Supreme Court galvanised opposition from individuals like Jeremy Riffkin and Pat Mooney and non-governmental organisations that continue to oppose patenting life in the USA and elsewhere.[99] Although Chakrabarty's patent claims were limited to genetically engineered mono-cellular living organisms and the Supreme Court in Chakrabarty never said that complex organisms were patentable, however, it took the position that the status of subject matter as living made no difference to its patentability. Therefore, a logical reading of Chakrabarty suggests that the Patent Act would seem to permit the patenting of any new, genetically engineered living organism.[100] Rebecca Eisenberg comments on the larger implications of Chakrabarty decision for patent jurisprudence as:

> As predicted by both proponents and opponents of patents on living organisms, investment in biotechnology R&D has flourished in the wake of *Diamond v. Chakrabarty*. But the full consequences of the expansive approach to patent eligibility endorsed by the *Chakrabarty* majority continue to be felt far beyond the biotechnology industry…A quarter century ago it was unclear whether the subject matter boundaries of the patent system were expansive enough to embrace biotechnology and information technology. Today, it is not clear whether the patent system has any subject matter boundaries at all.[101]

Courts in the USA started giving wide interpretation to patentable subject matter criterion under the US Patents Act to extend patent protection to higher life forms in the light of *Chakrabarty* decision.

**Plants as Patentable Subject Matter** In a 1985 case, *Ex parte Hibberd*,[102] the Board of Patent Appeals and Interferences reversed the USPTO's earlier rejection of a patent claiming corn plants and seeds as well as plant tissue cultures. The claims were related to plants produced through conventional cross-breeding but relied on new techniques such as cell culture and genetic analysis (but not recombinant DNA).[103] This case paved the way for patenting plants as by 1988, 42 patents on crop plants had been issued.[104]

**Animals as Patentable Subject Matter** In 1984, Standish Allen and Sandra Downing of the University of Washington and Jonathan Chaiton of the Coast Oyster

---

[98] Fowler (1994), 150 quoted in Dutfield, *supra* note 17, at 198.
[99] Dutfield, *supra* note 17, at 198.
[100] Demaine and Fellmeth (2002).
[101] Eisenberg (2006) quoted in Rimmer, *supra* note 42, at 44–45.
[102] *Ex parte Hibberd*, 227 U.S.P.Q. 443 (Bd. Pat. App. & Interferences 1985).
[103] Dutfield, *supra* note 17, at 198.
[104] *Ibid*.

Company applied for a patent on the production of the triploid-sterile Pacific oyster. The inventor's lawyer, David Maki, sought to extend the claim to include the triploid oyster itself.[105] The oyster had been genetically modified to demonstrate increased growth and, to be edible throughout all stages of its life cycle.[106]

The patent examiner rejected a number of claims in the application for lack of subject matter jurisdiction; by holding that polyploidy oyster was product of nature.[107] The examiner held that the animal produced by the method claimed was 'controlled by laws of nature and not a manufacture by man that is patentable'.[108] Further, the examiner had compared the claimed invention with the microorganisms claimed in *Diamond versus Chakrabarty* case and found that the claimed microorganisms were 'more akin to inanimate chemical compositions such as reactants, reagents, and catalysts than they are to horses and honeybees or raspberries and roses'.[109] Along with the lack of subject matter, the examiner also held that a number of claims were obvious in light of a previous publication. The previous publication recommended polyploidy as a way to increase growth in cultured oysters.[110]

On the later stage, the Board of Patents Appeals and Interferences pointed out that relevant test was not whether the subject matter was 'controlled by the law of nature', but whether it was naturally occurring.[111] The Board observed that '[t]he examiner has presented no evidence that the claimed polyploidy oysters occur naturally without the intervention of man, nor has the examiner urged that polyploidy oysters occur naturally'.[112] The Board decided that 'the claimed polyploidy oysters are non-naturally occurring manufactures or composition of matters within the confines of patentable subject matter under 35 USC 101'.[113] However, the Board agreed with the examiner's finding on obviousness of the claimed invention.[114] On appeal, the CAFC affirmed the decision of the Board of Patent Appeals and Interferences, without disturbing the finding that animals could be patentable subject matter.[115] The decision of the CAFC determined that Chakrabarty had opened the door to patents on genetic codes for multicellular animals that otherwise met the patentability requirements.[116]

In response to the decision of the Supreme Court of the USA in *Diamond versus Chakrabarty* and the ruling of the Board of Patent Appeals and Interferences in *Ex*

---

[105] Rimmer, *supra* note 42, at 84.
[106] Demaine and Fellmeth, *supra* note 100, at 318.
[107] *Ibid.*
[108] Rimmer, *supra* note 42, at 85, quoting *Ex Parte Allen* 2 USPQ 2d. P. 1425, 2) (1987).
[109] *Ibid.*
[110] Rimmer, *supra* note 42, at 85.
[111] Demaine and Fellmeth, *supra* note 100, at 319, quoting Ex parte Allen, 2 U.S.P.Q.2d (BNA) 1425 (Bd. Pat. Appeals & Interferences (1987).
[112] *Ex Parte Allen* 2 U.S.P.Q.2d,.1425, 2 (1987).
[113] *Ibid.*
[114] Rimmer, *supra* note 42, at 85.
[115] *Id.*, at 85–86.
[116] Demaine and Fellmeth, *supra* note 100, at 319.

## 2.2 Patentability of Biotechnology in the USA

*Parte Allen*, the Commissioner of the USPTO, Donald Quigg, released a notice, announcing that animals could constitute patentable subject matter. The notice states: '[t]he Patent and Trademark Office now considers non-naturally occurring non-human multicellular organisms including animals, to be patentable subject matter within the scope of 35 U.S.C. 101'.[117] The notice also made it clear that the Board's decision does not affect the principle and practice that products found in nature will not be considered to be a patentable subject matter under USC 101 and/or 102.[118] It ensures that a patent will not be granted to an article of manufacture or composition of matter occurring in nature unless given a new form, quality, properties or combination not present in the original article existing in nature in accordance with the existing law.[119]

The notice issued by USPTO attracted a lot of opposition from various farming and animal rights groups. They filed a lawsuit in Northern District of California, challenging the notice, which recognises animals as patentable subject matter.[120] The District Court dismissed the challenge on the first instance, mentioning that the notice was supported by relevant legal and administrative case law.[121] On appeal, the CAFC also dismissed the challenge from farming and animal groups. Nies C.J. ruled that the notice was consistent with the Supreme Court of the USA decision in *Diamond versus Chakrabarty*, and the USPTO rulings in *Ex parte Allen* and *Ex parte Hibberd*. He also made clear that the farmers, husbandry groups and animal rights organisations did not have standing to seek a declaration that animals are not patentable subject matter and an injunction against the issuance of animal patents.[122]

Following the CAFC's decision in *Ex Parte Allen*, the USPTO issued a patent on the Harvard Oncomouse in April 1988 to two genetic researchers, Philip Leder of Harvard Medical School and Timothy Stewart of San Francisco, who assigned it to the president and trustees of Harvard College.[123] Since the mouse was created for the study of breast cancer, therefore called oncomouse. The patent claim directed to the mouse itself or any mammal with the mouse's genetic idiosyncrasies. Accordingly, the patent claim covered the activated oncogene sequence in the animal's germ cells and somatic cells because the scope of the claim included the offspring of any mammal having the oncogene. To put it in other way, the claim on the mammal entailed a claim on at least the portion of the mammal's genome that coded for its novel morphology or physiology.[124] Following Harvard Oncomouse, a number

---

[117] United States Patent and Trademark Office (1987), 'Notice: animals-patentability', Official Gazette, United States Patent and trademark Office, 1077, 8, 21 April.

[118] *Ibid.*

[119] *Ibid.*

[120] *Animal legal Defense Fund v. Quigg 710 F.Supp. 728, 9 U.S.P.Q.2d 1816 (N.D.Cal. 1889); Animal Legal Defense Fund v. Quigg 932 F.2d 920 C.A. Fed. (Cal.), 1991.*

[121] *Animal legal Defense Fund v. Quigg 710 F.Supp. 728, 9 U.S.P.Q.2d 1816 (N.D.Cal. 1889).*

[122] Rimmer, *supra* note 42, at 87.

[123] *Id.*, at 90.

[124] Demaine and Fellmeth, *supra* note 100, at 318, citing U. S. Patent No. 4736,866 (issued Apr. 12, 1988).

of patents relating to animals had been issued and '[a]s of September 2003, about 454 animal patents has been granted in the USA, about half for disease models'.[125]

The Wisconsin Democrat Robert Kastenmeir introduced the *Transgenic Animal Patent Reform Bill 1989* (US) into the United States Congress. The Bill was aimed to provide patent defences for farmers in respect of the reproduction, use and sale of a patented transgenic farm animal and its offspring. Although the Bill was passed in the House of Representative, it was not debated in Senate before the end of Congress. In 1990, Kastenmeir had lost his seat in the 1990 Congress elections and the Bill was never reintroduced into the USA Congress.[126]

**Human Chimera and Humanoid: Testing the Extent of Patentable Subject Matter and Morality** In order to test the extent of patentable subject matter and ethics of patenting, Professor Stuart Newman of the New York Medical College and biotechnology opponent, Jermy Rifkin, made an announcement in 1997 that they would seek a patent on methods to create a chimera, a human-animal hybrid. This initiative was titled as 'the Human Chimera Patent Initiative'.[127] Afterwards, an application was filed in the USPTO seeking a patent on a technique combining human and animal embryonic cells to produce a single hybrid mouse-human embryo (the so called 'humouse').[128] 'The embryo could then be implanted into a human or animal surrogate mother to develop into a being of mixed human and animal composition—a not-so mythological chimera'.[129] The applicants had no intention to commercially exploit the humouse but to test the reaction of USPTO in this regard.[130] It was an attempt to force the USPTO to clear its position over the ethical dimensions of patent law and to promote public debate about the ethics of biotechnology, particularly with respect to animal research and human cloning.[131] Responding to this application, USPTO issued a media advisory by observing:

> The Patent and Trademark Office is required by law to keep all patent application in confidence until such time as a patent may be granted. However, the existence of a patent application directed to human/non-human chimera has recently been discussed in the news media. It is the position of the PTO that inventions directed to human/non-human chimera could, under certain circumstances, not be patentable because, among other things, they would fail to meet the public policy and morality aspects of the utility requirement.[132]

Consequently in 1999, the USPTO rejected the first application by Stuart Newman and Jeremy Rifkin because the invention 'embraces' a human being and failed the

---

[125] Dutfield, *supra* note 17, at 199, quoting Kelves (2006), 79.

[126] Rimmer, *supra* note 42, at 89.

[127] Rimmer, *supra* note 42, at 98–99.

[128] Demaine and Fellmeth, *supra* note 100, at 320.

[129] *Ibid.*

[130] *Ibid.*

[131] Rimmer, *supra* note 42, at 99.

[132] United states Patent and Trademark Office 'Media advisory: facts on patenting life forms having a relationship to humans', (1998) http://www.uspto.gov/web/offices/com/speeches/98-06.htm, quoted in Rimmer, supra note 42, at 99–100.

moral utility test.[133] The USPTO believed that the Congress did not intend 35 USC to include the patenting of human beings.[134] Moreover any claim 'directed to or including its scope a human being' would not be considered patentable because treating a human being as exclusive property is unconstitutional.[135] The USPTO also pointed out that the application did not meet the enablement and best mode criteria.[136]

Again in 2002, Newman and Rifkin refilled their patent application by arguing that the patent claims were not directed to a human being or human embryo, but rather a man-made chimeric animal developed from a chimeric embryo and the statute does not restrict patentability on the ground whether the claims embrace a human being.[137] However, the examiner at USPTO in 2004 rejected the application mentioning that the claimed invention was directed to non-statutory subject matter.[138] The examiner rejected the application for lack of specific and substantial utilities and not meeting the enablement and written description requirements.[139]

Stuart Newman claimed the decision of the USPTO as a moral victory and observed that the action of the USPTO reflected the absence of an effective criterion in the patent system to determine the issue of relative humanity of a genetically engineered organism:

> But if you could genetically engineer the chimera so that the human component will be a known percentage of the organism then the USPTO might be better satisfied. I don't think that the rejection of this patent will impede research in the field. I do hope however, that it stimulates legislative guidelines. With commercial incentive alone it is only a matter of time before such an organism is made.[140]

The human chimera episode made it clear that though the US patent law does not contain morality and public order provisions as grounds for rejecting a patent, however, it recognises its utility with regard to few serious issues. This concern has been echoed in an editorial in Nature Biotechnology, which noted that 'no country's patent system has yet found a way of extricating itself from the philosophical and political morass associated with patent applications that encroach on definitions of humanness'.[141] Recently, the approach adopted by USPTO towards patenting

---

[133] Demaine and Fellmeth, *supra* note 100, at 320.

[134] Rimmer, *supra* note 42, at 100.

[135] Demaine and Fellmeth, *supra* note 100, at 320, quoting Donald J. Quigg, Statement by Assistant Sec'y of Comm. & Comm'r of Pat. & Trademarks, 1077 Off. Gazette U.S. Pat. &Trademark OFF. 24 (Apr.7, 1987) at 24.

[136] Rimmer, *supra* note 42, at 100.

[137] *Ibid.*

[138] *Ibid.*

[139] *Id.*, at 101–102.

[140] *Id.*, at 102, quoting Editorial, 'Hybrid too human patent: case highlights lack of criterion for genetically modified organisms', *Nature Review Drug Discovery*, (2005), http://www.nature.com/news/2005/050328/full/nrd1710.html, 31 March.

[141] *Id.*, at 102 quoting Editorial, 'Patenting pieces of people', 21 *Nature Biotechnology*, 341, (1 April, 2003).

of human organism has been given the backing of law, by Leahy-Smith America Invents Act 2011 (AIA). Sec. 33(a) of the said Act reads: 'Notwithstanding any other provision of law, no patent may issue on a claim directed to or encompassing a human organism'.[142] This provision of the Leahy-Smith America Invents Act is in consonance with the long-standing policy of the USPTO that a claim encompassing a human being is not patentable, and the said Act does not bring any new change in the existing patent practice. However, it recognises the existing practice by conferring the effect of law. Sec. 33(b) mentions that Section 33(a) shall apply to patent applications filed on or after the date of enactment (i.e. 16 September 2011) of the Leahy-Smith America Invents Act 2011 and to the applications pending on the said date.[143] The words 'directed to' and 'encompassing' are not very clear and open for interpretation. It is to be seen how it would be interpreted in biotech patent cases.

**Human Gene as Patentable Subject Matter** Human genes have become a common subject matter for patents following Chakrabarty decision. One of the most common objections against the gene patents is that genes are naturally occurring entities that are there to be discovered but not invented. In the USA, although the patent statute states that both discoveries and inventions qualify as patentable subject matter, however, in practice the law does not permit the patenting of natural phenomena.[144]

The USPTO and courts in the USA have long recognised that isolated and purified substances do not exist in nature, hence patentable.[145] This has been the general trend of the USPTO and courts regarding chemical products. This analogy has been applied to human DNA also by considering it a chemical. Before Chakrabarty, there were several cases in which the USPTO and courts recognised that isolated and purified chemical substances were patentable. First cited case in this series was *Park-Davis & Co. versus H.K. Mulford & Co.*,[146] where the applicant had patented adrenalin. The claim of the applicant states: A substance possessing the herein-described physiological characteristics and reactions of the suprarenal glands in a stable and concentrated form, and practically free from inert and associated gland tissue.[147] The court held that a substance derived and purified from nature could be patentable.[148] In a 1970 case, *In re Bergstrom*,[149] the inventors claimed 'naturally occurring', the prostaglandin compounds, PGE2 and PGE3, that they had extracted and purified from prostate gland. The court held that the 'sufficiently pure'

---

[142] Sec. 33(a), Leahy-Smith America Invents Act (AIA) 2011, available at http://www.gpo.gov/fdsys/pkg/BILLS-112hr1249enr/pdf/BILLS-112hr1249enr.pdf (Visited on Sept. 27, 2011).

[143] Sec. 33(b), Leahy-Smith America Invents Act (AIA) 2011, available at http://www.gpo.gov/fdsys/pkg/BILLS-112hr1249enr/pdf/BILLS-112hr1249enr.pdf (Visited on Sept. 27, 2011).

[144] *Diamond v. Diehr, 450 U.S. 175, 185 (1981)*.

[145] Goldstein and Gold (2002).

[146] 196 F.496 (2nd Cir. 1912).

[147] *Id.* 497.

[148] *Id.*, at 498.

[149] 427 F.2d 1394 (C.C.P.A. 1970).

## 2.2 Patentability of Biotechnology in the USA

prostaglandins did not exist in nature and ruled these to be patentable.[150] In another case *In re Bergy*,[151] the applicant claimed a culture of a naturally occurring bacterium that produced an antibiotic. 7 *In re Bergy*, 563 F.2d 1031 (C.C.P.A. 1977). The applicant's claim related to 'a… biologically pure culture of the microorganism *Streptomyces velosus*'.[152] Here, the court (CCPA) again held that 'biologically pure cultures' of the microorganisms did not exist in nature and could be patentable.[153] In Chakrabarty, as mentioned earlier, the US Supreme Court held that a human made non-natural microorganism was patentable.[154] Since a genetically modified microorganism such as that of Chakrabarty was not even a product of nature, the legal analysis was simpler than in Bergy.[155]

Applying the chemical analogy to biotechnology and following the same logic of isolated and purified natural substances, USPTO started issuing patents to isolated or purified human genes encoding protein drugs,[156] diagnostic probes,[157] receptor immunogens and gene replacement therapies.[158] Almost all claims relating to gene patents use the word 'isolated' and 'purified' in order to be consistent with the adrenalin and prostaglandin precedents.[159] Here, it becomes pertinent to explain what the terms isolated and purified connote. The Utility Examination Guidelines interpret the word 'isolation' to mean separation of DNA 'from its natural state'.[160] Generally, 'purified' means excluded from the way the particular DNA occurs in nature.[161] However, the term 'purified' does not have an exact and identical definition in all circumstances. Applicants are encouraged to define the term in the patent specification for each case.[162]

---

[150] *Id.*, at 1401–02.

[151] 563 F.2d 1031 (C.C.P.A. 1977).

[152] *Id*, at 1032.

[153] *Id.*, at 1035.

[154] 447 U.S. 303 (1980).

[155] Goldstein and Gold, *supra* note 145.

[156] *Id.*, at 1316–17; Genentech Inc. received a patent for a DNA isolate consisting essentially of a DNA encoding human tissue plasminogen activator (1988) (U.S. Patent No. 4,766,075); Kiren Amgen obtained a patent (1987) for "purified and isolated" DNA sequences encoding erythropoietin" (U.S. Patent No. 4,703,008).

[157] *Id*, at 1317; OncorMed has obtained a patent (1988) for an isolated coding sequence of the BRCA1 gene (U.S. Patent No. 5,750,400), which can be used for screening individuals with an increased genetic susceptibility to breast or ovarian cancer because of the inherited mutation of the BRCA1 gene.

[158] *Ibid;* Human Genome Sciences, Inc., has obtained a patent (2001) for polynucleotides encoding human tr10 receptor, a member of the tumor necrosis factor (TNF) receptor superfamily and the TRAIL receptor subfamily. (U.S. Patent No. 6,214,580).

[159] *Ibid.*

[160] *Ibid.*, quoting *Utility Examination Guidelines*, 6 Fed. Reg. 1092, 1093 (Jan. 5, 2001).

[161] *Ibid.*, citing Biotechnology: *Patents, Licensing & FDA Practice* at I-48, Patent Resources Group, Inc. (2000).

[162] *Ibid.*, citing Biotechnology: *Patents, Licensing & FDA Practice* at I-49, I-50. Patent Resources Group, Inc. (2000).

Although USPTO had been issuing patents on genes considering it as chemicals, however, it was in 1991, when in the case of *Amgen Inc versus Chugai Pharma. Co.*,[163] the CAFC specifically held that 'a gene is a chemical molecule, albeit a complex one'.[164] In Amgen, three companies fought over the patent rights of the DNA sequences that encode the human erythropoietin (epo) protein, which stimulates the production of red blood cells. epo has a significant role in many doping scandals due to its potential to enhance the blood's ability to carry oxygen.[165] The Amgen's product claim, upheld by the Federal Circuit was considerably broad. The claim read as follows: 'A purified and isolated DNA sequence, encoding human erythropoietin'.[166] Here, Amgen's claim extended to the purified and isolated form of any gene that coded for the epo protein. In the context of Amgen's patent claim, the terms 'purified and isolated' meant that the coding region of the relevant DNA had been reproduced outside its natural environment.[167]

By 1991, the Federal Circuit has acquiesced in the proposition that the words 'purified and isolated' were sufficient to distinguish a claimed gene from its naturally occurring counterpart.[168] The essence of the Amgen's invention was 'the successful cloning of the epo gene'.[169] A clone, or cDNA version of a naturally occurring gene, differs from the naturally-occurring variant in that introns, or non-coding regions, are absent.[170] This decision has opened the floodgates for human gene patents and much of this gene patent rush spurred from the rapidly evolving capabilities for gene sequencing and cloning genes.[171]

The first generation of patents on DNA sequences as in the *Amgen* case from the 1980s were directed towards gene encoding therapeutically significant proteins. The patent claims on these genes were directed to isolated and purified DNA sequences encoding these proteins (generally, cDNA molecules created by reverse transcription), a recombinant vector that includes the DNA sequences, and transformed host cells that include the vectors. Each of these claims covered tangible materials used to make pharmaceutical products having the similar effect to a patent on drug.[172] In the USA, the USPTO and the courts considered these patents analogous to patents on new chemical compounds. 'The analogy may never have been perfect but it worked in the sense of motivating investments in the biotechnology industry and in the development of the new products'.[173]

---

[163] *Amgen Inc. v Chugai Pharma. Co. 927 F.2d 1200 (Fed. Cir. 1991)*.

[164] *Id.*, at 1206.

[165] Conley (2009).

[166] *Amgen*, supra note 163, at 1204, quoting U.S. Patent No. 4, 703,008 (Filed Nov. 30, 1984).

[167] Conley, *supra* note 165, at 115–16.

[168] *Id.*, at 116.

[169] *Amgen Inc. v. Chugai Pharma. Co., 13 U.S.P.Q.2d 1737, 1759 (D. Mass. 1989),* quoting *Diamond v. Chakrabarty, 447 U.S. 303, 309 (1980).*

[170] Conley, *supra* note 165, at 116.

[171] Nese (2009).

[172] Eisenberg (2003).

[173] Ibid.

## 2.2 Patentability of Biotechnology in the USA

In recent years, there has been a paradigm shift in the gene patent practice. The development of new tools and techniques for detecting genetic differences among individuals enabled researchers to bypass the stages of protein isolation and characterization (prevalent in the case of first generation gene patents) and to identify directly the genes associated with diseases (or disease susceptibilities) through positional cloning.[174]

This heralded a new era of genetic discoveries having immediate value as diagnostic products. These discoveries were also useful as research tools in the search of diagnostic products, but the relationship between gene and therapeutic product was typically less straightforward than it had been for the first generation of biotechnology products. Though patents on these discoveries were similar in form to patents on genes encoding therapeutic protein, however, they performed a less familiar role in biomedical community raising conflicting concerns for people and institutions who had barely taken note of the first generation of gene patents.[175] For instance, professional societies of doctors and clinical geneticists became critics of disease gene patents as well as exclusive licenses to perform DNA diagnostic tests.[176]

Further, gene has become more important as information than a physical entity with the advancements in genomics. The advent of high throughput DNA sequencing unleashed a large amount of DNA sequence information in advance without fully understanding the function of particular gene sequence.[177] This has led to patenting of gene and gene fragments with little or no utility.

This was reflected in an attempt made by the National Institutes of Health (NIH) to obtain patents on ESTs in 1991. 'Unlike fully sequenced genes with known and exploitable functions, an EST indicates nothing about the full sequence or function of the transcribed nucleic acid with which it is associated'.[178] The NIH's filing of patent application claiming ESTs had provoked great controversy among biomedical researchers, including many working on the Human Genome Project because of the perception that broad EST patents would inhibit research on gene function. This led to James Watson's resignation as the director of the Genome Project in 1992 over patenting dispute. However, the NIH later reversed its position on EST patents and abandoned its applications. Nevertheless, other organisations filed similar patent applications in the 1990s and maintained those filings.[179]

Commenting on the actual value of ESTs, Rebecca Eisenberg noted:

> The most obvious value of ESTs was not the speculative value of particular gene fragments for therapeutic or diagnostic uses, but the immediate value of growing collections of such sequences for use in gene discovery. With this shift, patenting genes started to look less like patenting end products and more like patenting scientific information.[180]

---

[174] Eisenberg (2002).
[175] Merrill and Mazza (2006b).
[176] *Ibid.*
[177] *Ibid.*
[178] Davis et al. (2005).
[179] *Ibid.*
[180] Eisenberg, *supra* note 174, at 1383.

Scientists opposed this move by arguing that 'the progress of biomedical research would be better served by making the human genome freely available than permitting its balkanisation through patent claims and restrictive licensing agreements'.[181] The shift of EST sequencing from NIH to the private sector made the opposition more vehement.[182]

After continued gene patent rush, the USPTO began receiving heavy criticism for granting patents too liberally. In response, the USPTO issued utility guidelines in 2001, tightening the utility criterion by prescribing specific, substantial and credible utility.[183] Despite 2001 guidelines, gene patent rush continued and as of 2005 '[n]early 30, 000 human genes have been patented' in the USA.[184] In 2005, in the case of *In re Fisher*,[185] the CAFC rejected a claim on ESTs for the lack of utility in the light of 2001 Guidelines,[186] however, it has not ruled out the possibility of ESTs as patentable subject matter.

Apart from making the concern regarding patenting of research tools, the EST controversy also highlighted the character of genomic discoveries as information, as distinguished from tangible molecules. This is due to the fact that much of the value ESTs lay in databases, rather than in tangible materials stored in a wet laboratory.[187] Gene and gene fragments as information remain problematic candidates to pass the patent eligibility test because they come under the exceptions, scientific truths and abstract ideas, long recognised by courts in several old cases.[188] Although the courts and USPTO initially resisted the extension of the patent protection to information technology, however, they gradually allowed the same. The overall trend of the CAFC is towards the expansive interpretation of the scope of patent eligible subject matter-even for categories of inventions that prior decisions seemed to exclude from the protection of the patent statute in order to make the patent system 'responsive to the needs of the modern world'.[189]

The 1998 decision in *Street Bank & Trust versus Signature Financial Group*[190] reflected this trend by upholding the patentability of a computer-implemented accounting system for managing the flow of funds in partnerships of mutual funds that pool their assets.[191] It was contended that this invention fell under some previous

---

[181] *Ibid.*

[182] *Ibid.*

[183] USPTO Utility Examination Guidelines, 66 Fed. Reg. 1092, 5 Jan 2001.

[184] Nese, *supra* note 171, at 153, quoting John Merz and Mildred Cho, "What Are Gene Patents and Why are People Worried About Them?" 8 *Community Genetics* 203 (2005).

[185] *In re Fisher*, 421 F.3d 1365 (C.A., Fed., 2005).

[186] *Supra* note 183.

[187] Merrill and Mazza, *supra* note 175, at 73.

[188] *Id.*, at 73, citing *Leroy v. Tatham*, 55 U.S. 156, 175 (1853); *Mackay Co. v. Radio Corp.*, 306 U.S. 86, 94 (1939); *Gottschalk v. Benson*, 409 U.S. 63 (1972); *Diamond v. Chakrabarty* 447 U.S. 303 (1980) and *Dickey-John Corp. v. Tapetronics Corp.*, 710 F.2d 329 (7th Cir. 1983).

[189] Merrill and Mazza, *supra* note 175, at 77.

[190] *Street Bank & Trust Co. v. Signature Financial Group, Inc., 149 F.3d 1368 (Fed. Cir. 1998).*

[191] Merrill and Mazza, *supra* note 175, at 77.

## 2.2 Patentability of Biotechnology in the USA

judicial limitations that arguably excluded mathematical algorithms[192] and business methods[193] from patent protection.[194] Responding to the objection regarding mathematical algorithms, the CAFC held that it excluded from patent protection only 'abstract ideas constituting disembodied concepts or truths that are not "useful"'.[195] In response to the second objection relating to business method, the CAFC explained that '[t]he business method exception has never been invoked by this court, or [its predecessor], to deem an invention unpatentable'. and that other courts that had appeared to apply the business method exception always had other grounds for arriving at the same decision.[196] The CAFC established the principle that an invention is patent eligible, if it produces a 'useful, concrete and tangible result'.[197]

In another recent case, *AT & T Corp. versus Excel Communications, Inc.*,[198] the court declined to focus upon the 'physical limitations inquiry', which had been used to distinguish between unpatentable mathematical algorithms and patentable computer-implemented inventions in prior decisions.[199] The court, on the other hand, focused on 'whether the mathematical algorithm is applied in a practical manner to produce a useful result'.[200] This indicates the merger of the patent eligibility issue with the issue of utility. It is, however, not clear whether the court would allow patent claims to information itself as long as it is useful. There has been no clear indication of tightening of the patentable subject matter standard as there have not been any court decisions which struck down gene or protein patents on subject matter grounds. However, there are some indirect hints that change may be in the air.[201]

**Tightening of Subject Matter Boundary: Revisiting Patentable Subject Matter Test in the Light of Diagnostic Gene Patents** There was some hint regarding tightening of patentable subject matter boundary in *Laboratory Corp. Of America Holdings versus Metabolite Laboratories, Inc.*,[202] where the Supreme Court agreed to hear an appeal of Metabolite to specifically address the issue whether the patent violated the court's prohibition against patenting 'laws of nature, natural phenomena, and abstract ideas'.[203] However, after party briefs and arguments, the majority

---

[192] *Gottschalk v. Bensen, 409 U.S. 63 (1972); Parker v. Flook, 437 U.S. 584 (1978).*

[193] *Hotel Security Checking Co., v. Lorraine Co., 160 F.467 (2d Cir. 1908).*

[194] Merrill and Mazza, *supra* note 175, at 77.

[195] *Street Bank*, supra note 190, at 1373.

[196] *Id.*, at 1375–76, quoted in Merrill and Mazza, *supra* note 175, at 77–78.

[197] *Id.*, at 1373.

[198] *AT & T Corp. v. Excel Communications, Inc.*, 172 F.3d 1352 (Fed. Cir. 1999).

[199] Merrill and Mazza, *supra* note 175, at 79.

[200] *Ibid.*, quoting *supra* note 198, at 1358.

[201] Conley, *supra* note 165, at 120.

[202] *Laboratory Corp. Of America Holdings v. Metabolite Laboratories, Inc.*, 126 S. Ct. 2921 (2006).

[203] Holman (2007).

decided not to decide the case because the issue of patentable subject matter had not been directly addressed in the lower courts.[204] The case involved:

> [A] patent that claims a process for helping to diagnose deficiencies of two vitamins, folate and cobalmin. The process consists of using any test (whether patented or unpatented) to measure the level in a body fluid of an amino acid called homocysteine and then noticing whether its level is elevated above the norm; if so, a vitamin deficiency is likely.[205]

Both, the trial court and the CAFC decided that the patent was valid and the Laboratory Corporation of America (LabCorp) was liable for inducing infringement when it encouraged doctors to order diagnostic tests for measuring homocysteine.[206] Though LabCorp had defended the case on various grounds including that the patent was invalid for being overbroad; however, it had failed to raise the issue of patentable subject matter explicitly in the lower courts.[207]

In its successful petition for certiorari, LabCorp raised the following question:

> Whether a method patent [setting forth an indefinite, undescribed and non-enabling step) directing a party simply to "correlate" test results can validly claim monopoly over a basic scientific relationship [used in medical treatment] such that any doctor necessarily infringes the patent merely by thinking about the relationship after looking at a test result.[208]

The court, however, dismissed the writ as improvidently granted. The majority ruled that the earlier decision by the court to hear the case had been a mistake, essentially because the issue of patentable subject matter had not been directly addressed in the lower courts, and because the Supreme Court normally does not decide issues that were not already argued in the lower court.[209] However, a vocal minority comprising Justice Breyer, joined by Justices Stevens and Souter, would have decided the case and invalidated the patent claim.[210] The minority expressed strong concerns regarding the policy implications of such patents and shown its willingness to decide the case in such a way that would restrict the patentability of processes that in effect embody a law of nature.[211]

The issue of patentable subject matter was specifically raised in *In re Bilski*[212] case. In the said case, the CAFC had to decide whether certain process claims constitute patent eligible subject matter under Section 101. The case involved the patentability of a process to calculate the hedging of risks between commodities and market participants. Claim I of Bilski's application recites:

---

[204] *Ibid.*

[205] *Lab. Corp.*, *supra* note 202 (Breyer J. dissenting).

[206] *Ibid.*

[207] *Id.*, at 2925 (LabCorp did not refer in the lower courts to Sec. 101 which sets forth the subject matter that is patentable, and within the bounds of which the 'law of nature' principle most comfortably fits.) cited in Conley, *supra* note 165, at 121.

[208] *Ibid.* (emphasis added; citation omitted) in Conley, *supra* note 165, at 121.

[209] Holman, *supra* note 203.

[210] *Ibid.*

[211] Ibid.

[212] *In re Bilski*, 545 F.3d 943 (Fed. Cir. 2008) (*en banc*).

## 2.2 Patentability of Biotechnology in the USA

> A method for managing the consumption of risk costs of a commodity [,] comprising the steps of:
> 1. Initiating a series of transactions between [a] commodity provider and consumers of said commodity.
> 2. Identifying market participants for said commodity....;
> 3. Initiating a series of transactions between said commodity provider and said market participants... such that said series of market participant transactions balances the risk position of said series of consumer transactions.[213]

In this case, the CAFC evolved a new standard as the 'machine or transformation' test to determine whether the claimed method constituted a patent-eligible process under Section 101 of the Patent Act. According to the court, a claim contains patent-eligible subject matter if '(1) it is tied to a particular machine or apparatus or (2) it transforms a particular article into a different state or thing'.[214] The purpose behind the evolution of machine or transformation test was to prevent the patenting (or pre-emption) of 'fundamental principles', such as abstract ideas, laws of nature, or other traditionally-ineligible subject matter.[215]

While rejecting the process claim in *Bilski*, the court concluded that since Bilski's process claim was not tied to any specific machine or apparatus and did not transform any article to a different state or thing, therefore, failed the machine and transformation test.[216] However, the court did not explain what does and does not satisfy the transformation prong and leaved such determination on future cases. In the present case, the court opined that a process would satisfy this prong (and thus constitute eligible subject matter) if it transformed 'an article into a different state or thing'.[217] Emphasising the importance of transformation prong, the court maintained that '[t]his transformation must be central to the purpose of the claimed process'.[218]

Considering the question as to what sort of things constituted articles, the court concluded that mere abstract concepts or traditionally unpatentable subject matter did not constitute an article for these purposes.[219] The court observed that in Bilski, the claimed process to calculate the hedging of risks between commodities and market participants was, in fact, the use of abstract ideas and mental process. Quoting an earlier case, the court concluded that a claim that requires an 'application of [only] human intelligence to the solution of practical problems' is merely a claim to a fundamental principle, and thus unpatentable.[220]

Similar to Bilski's business method claim, diagnostic gene patents 'seek to claim a non transformative process that encompasses [a fundamental principle] without

---

[213] *Ibid.*
[214] *Id.*, at 954.
[215] *Id.*, at 963.
[216] *Ibid.*
[217] *Id.*, at 962.
[218] *Ibid.*
[219] *Ibid.*
[220] Id, at 965 (quoting *In re Comisky*, 499 F.3d 1365, 1377–79 (Fed. Cir. 2007)).

the aid of any other device'.²²¹ Therefore, a diagnostic gene patent claim, which simply involves comparing a particular gene or group of genes with a sample of DNA to determine the presence or absence of gene or genes, fails the 'machine-or-transformation test'. The reason for such failure is that 'the claim recites no physical or chemical alteration; nor does it refer to anything changing, reforming, or otherwise becoming another thing'.²²²

On appeal, the case reached to the Supreme Court and the court held that the machine-or-transformation test is not the sole test for patent eligibility under Section 101. The court maintained that the test may be a useful and important clue or investigative tool; it is not the sole test for deciding whether an invention is a patent-eligible 'process' under Section 101.²²³ The court criticized the CAFC's decision by saying that the CAFC violated two principles of statutory interpretation: Courts "should not read into the patent laws limitations and conditions which the legislature has not expressed,' '²²⁴ and, '[u]nless otherwise defined, "words will be interpreted as taking their ordinary, contemporary, common meaning"'.²²⁵

Expressing the ill effects of machine or transformation in the information age, the court observed:

> The machine-or-transformation test may well provide a sufficient basis for evaluating processes similar to those in the Industrial Age—for example, inventions grounded in a physical or other tangible form. But there are reasons to doubt whether the test should be the sole criterion for determining the patentability of inventions in the Information Age. As numerous amicus briefs argue, the machine-or-transformation test would create uncertainty as to the patentability of software, advanced diagnostic medicine techniques, and inventions based on linear programming, data compression, and the manipulation of digital signals.²²⁶

*Association for Molecular Pathology versus USPTO* The *Association for Molecular Pathology et al. versus United States Patent and Trademark Office et al.*²²⁷ involved a law suit challenging the legality of patents on breast cancer genes BRCA1/2 held by Myriad Genetics and the University of Utah. The American Civil Liberties Union (ACLU) and Public Patent Foundation (PPF) represented the plaintiffs, and Jones Day represented Myriad. The case was heard in the Southern District of New York by Judge Sweet and decided on 29 March 2010. In his decision, Judge Sweet divided Myriad's patent claims into two groups:

1. Patents claiming gene sequences that had been isolated from DNA and;
2. Patents claiming methods for comparing and analyzing gene sequences to identify the presence of mutations corresponding to a predisposition to breast or ovarian cancer.²²⁸

---

²²¹ *Id.*, at 965.
²²² Nese, *supra* note 171, at 171.
²²³ *Bilski et al. V. Kappos*, No. 08–964 (2010).
²²⁴ *Diamond v. Diehr*, 450 U. S. 175, 182 quoted in *Bilski, supra* note 223, at 2.
²²⁵ *Ibid.*
²²⁶ *Bilsk*i, *supra* note 223, slip op. at 9.
²²⁷ *Association for Molecular Pathology et al. v United States Patent and Trademark Office et al.*, No. 09-CV 4515, (S.D.N.Y. March 29, 2010).
²²⁸ *Id.*, at 2.

## 2.2 Patentability of Biotechnology in the USA

Judge Sweet rejected both sets of patents under section 101 of the Patent Act by holding that isolated genes are insufficiently different from naturally occurring DNA and, thus, are ineligible for patent protection.[229] He also maintained that comparisons of DNA sequences involved in these patents are abstract mental process, therefore also unpatentable.[230]

Judge Sweet held that Myriad's isolated genes are insufficiently different from naturally occurring genes in the body and falls under the product of nature exception. He emphasised on the separation of patentable subject matter from other patent-eligibility criteria-novelty, utility and non-obviousness. The court maintained that patentable subject matter is a free-standing inquiry separate from novelty, utility and non-obviousness, and one cannot prove that something is patent-eligible human made invention as opposed to a product of nature by simply showing that it is novel and useful. Rather, one has to show separately, whether something is patentable subject matter or not.[231]

Myriad contended that the isolated gene were not simply a product of nature by arguing that the genes were 'substantially separated from other cellular components with naturally accompany n native human sequence [such as] human genome sequences and proteins'.[232] Responding to this contention, Judge Sweet concluded that 'purification of a product of nature, without more, cannot transform into patentable subject matter. Rather, the purified product must possess "markedly different characteristics" in order to satisfy the requirements'.[233] Differentiating DNA from a chemical molecule, Judge Sweet observed:

> DNA represents the physical embodiment of biological information, distinct in its essential characteristics from any other chemical found in nature. It is concluded that DNA's existence in an "isolated" form alters neither this fundamental quality as it exists in the body nor the information it encodes.[234]

The court further added:

> In light of DNA's unique qualities as a physical embodiment of information, none of the structural and functional differences cited by Myriad between native BRCA1/2 DNA and the isolated BRCA1/2 DNA claimed in the patents-in-suit render the claimed DNA "markedly different." This conclusion is driven by the overriding importance of DNA's nucleotide sequence to both its natural biological function as well as the utility associated with DNA in its isolated form. The preservation of this defining characteristic of DNA in its native and isolated forms mandates the conclusion that the challenged composition claims are directed to unpatentable products of nature.[235]

The court's stand that genes are distinct from chemicals is a radical departure from precedent, as in *Amgen Inc. versus Chugai Pharm. Co.*, the court held that '[a] gene

---

[229] *Id.*, at 3–4.
[230] *Ibid.*
[231] *AMP v USPTO, supra note 227*, at 99–102.
[232] *Id.*, at 92.
[233] *Id.*, at 121.
[234] *Id.*, at 3–4.
[235] *Id.*, at 125.

is a chemical compound, albeit a complex one'.[236] This demolishes an established practice of USPTO of granting patents on isolated and purified genes. In the light of this decision, genes now fall outside of the *In re Bergstrom*[237] holding in which a purified form of a compound was held patentable, even though its impure form was known.[238]

Regarding second set of Myriad's method claims, the court invoked the 'machine or transformation' test propounded by the CAFC's in *In re Bilski*[239] case. Applying the test, Judge Sweet concluded that none of the patent claims were tied to any particular machine, nor did they bring about a tangible transformation of anything. He explained that 'because the claimed comparisons of DNA sequences are abstract mental processes, they also constitute unpatentable subject matter'.[240] Judge Sweet further added that even if the claims were construed in such a way that they constituted 'physical transformations associated with isolating and sequencing DNA, they would still fail the "machine or transformation" test under Section 101 for subject matter patentability'.[241]

Myriad Genetics Inc. and the Directors of the University of Utah Research Foundation made an appeal from the decision of the US District Court for the Southern District Court of New York to the CAFC. The CAFC reversed the District Court's decision that Myriad's composition claims to 'isolated' DNA molecules cover patent-ineligible products of nature under Section 101 since the molecules as claimed do not exist in nature. The court also reversed the District Court decision that Myriad's method claim to screening potential therapeutics via changes in cell growth rates is directed to a patent-ineligible scientific principle. However, the court affirmed the District Court's decision that Myriad's method claims directed to 'comparing' or 'analysing' DNA sequences are patent ineligible since such claims include no transformative steps and cover only patent ineligible abstract mental steps.[242]

Judge Louire wrote the opinion of the court by announcing the judgement and providing the rationale. Judge Moore wrote a concurring opinion by joining all aspects of the judgement. Although, she agreed with Judge Lourie's reasoning with respect to the method claims and the patentability of isolated cDNA sequences, however, she had a slightly different reason for upholding the patentability of *DNA* sequences. Judge Bryson, on the other hand, joined in the judgement with respect to the method claims and the patentability of longer sequences of cDNA. However, he voted against the patentability of all isolated DNA sequences as well as very short

---

[236] *Amgen, supra* note 163, at 1206.

[237] *In re Bergstrom*, 427 F.2d 1394 (C.C.P.A. 1970).

[238] *Id.*, at 1401–02.

[239] *Bilski, supra* note 212, at 954.

[240] *AMP v USPTO, supra* note 227, at 4.

[241] *Id.*, at 147.

[242] *Association of Molecular Pathology et al. v. United States Patent and Trademark Office et al., available at* http://www.cafc.uscourts.gov/images/stories/opinions-orders/10-1406.pdf,2011) (Visited on August 30, 2011).

cDNA sequences, and would thus have affirmed the District Court on that specific point.[243]

The most notable outcome of the CAFC decision was that it explicitly invalidated the comparing and analysing method claims which have been notoriously exercised by Myriad. It also provided certainty to the long-standing practice of the USPTO of granting patents on isolated and purified genes.

Meanwhile, another case which had a great impact on diagnostic gene patents was *Prometheus Labs, Inc. versus Mayo Collaborative Servs.*[244] On 20 June 2011, the US Supreme Court granted Mayo Medical Laboratories a writ of certiorari in its appeal from the CAFC that upheld two diagnostic method patents held by Prometheus Laboratories.[245]

In Prometheus, patents at issue involved a method of administering thiopurine drugs (used in the treatment of gastrointestinal and other autoimmune diseases); measuring the drug level in the patient's body and then adjusting the doses of drug.[246] Prometheus manufactured a testing kit, previously used by defendants Mayo Collaborative Services and the Mayo Clinic Researcher.[247] Later on, Mayo planned to begin using its own kit, testing for the same metabolites but using different levels to determine toxicity.[248] Eventually, Prometheus sued Mayo for patent infringement, prompting Mayo to suspend its plans pending resolution of the case.[249] In March 2008, the California District Court invalidated a pair of patents (US Patent No. 6,355,623 and No. 6,680,302) exclusively licensed to Prometheus Laboratories, holding that the claimed inventions were not patentable under Section 101 of the Patent Act.[250]

The CAFC had overturned the ruling by the District Court on 16 September 2009.[251] The CAFC held that the claimed methods satisfied the machine or transformation test, which was used to decide whether a particular method qualify as patentable subject matter under Section 101. The court specifically held that the administration and measurement steps worked a sufficient transformation of the body to satisfy the 'method or transformation' (MOT) test.[252]

Mayo applied for certiorari to the Supreme Court. In the meantime, the Supreme Court took up the appropriateness of the MOT test for method patents in *Bilski versus Kappos*.[253] Limiting the scope of MOT in *Bilski versus Kappos*, the Supreme

---

[243] Conley and Vorhaus (2011a).
[244] *Prometheus Labs, Inc. v. Mayo Collaborative Servs, 581 F.3d 1336 (Fed. Cir. 2009).*
[245] *Id.*, at 1339.
[246] *Id.*, at 1339–40.
[247] *Id.*, at 1340.
[248] *Ibid.*
[249] *Ibid.*
[250] *Prometheus Labs, Inc. v. Mayo Collaborative Servs; No. 04-CV-1200 JAH (RBB), 2008 WL 878910, at 6 (S.D. Cal. Mar. 28, 2008), at 14.*
[251] *Prometheus, supra* note 244, at 1350.
[252] Conley and Vorhaus (2011b).
[253] *Bilski, supra* note 223.

Court held that MOT was not the exclusive test for evaluating method patents[254] but failed to issue a broader guidance including biotechnology industry leaving great confusion regarding gene patents.[255]

Almost immediately after the decision of the Supreme Court in *Bilski versus Kappos*, the Supreme Court granted certiorari in *Prometheus* vacating the CAFC decision and ordering that the case be considered by the CAFC in light of *Bilski versus Kappos* (remand).[256] The CAFC reconsidered *Prometheus* and reached to its previous conclusion that '623' and '302'patents are valid under Section 101 in light of *Bilski versus Kappos*. In the second certiorari appeal, Mayo has been successful in getting the Supreme Court to hear the case.[257] After hearing the case, the Supreme Court reversed the decision of CAFC and decided unanimously that Prometheus' process was not patentable. The court reasoned:

> Because the laws of nature recited by Prometheus' patent claims—the relationships between concentrations of certain metabolites in the blood and the likelihood that a thiopurine drug dosage will prove ineffective or cause harm—are not themselves patentable, the claimed processes are not patentable unless they have additional features that provide practical assurance that the processes are genuine applications of those laws rather than drafting efforts designed to monopolize the correlations. The three additional steps in the claimed processes here are not themselves natural laws but neither are they sufficient to transform the nature of the claims…[258]

Therefore, the court made it clear that claims directed to a method of giving a drug to a patient, measuring metabolites of that drug, and with a known threshold for efficacy in mind, deciding whether to increase or decrease the dosage of the drug, were not patent eligible subject matter.[259]

On 13 June 2013, the Supreme Court in *Association for Molecular Pathology versus Myriad Genetics*[260] held that '[a] naturally occurring DNA segment is a product of nature and not patent eligible merely because it has been isolated, but cDNA is patent eligible because it is not naturally occurring'.[261] As regards to the patentability of cDNA, the court opined:

> cDNA is not a "product of nature," so it is patent eligible under § 101. cDNA does not present the same obstacles to patentability as naturally occurring, isolated DNA segments. Its creation results in an exons-only molecule, which is not naturally occurring. Its order of the exons may be dictated by nature, but the lab technician unquestionably creates something new when introns are removed from a DNA sequence to make cDNA.[262]

---

[254] *Ibid.*

[255] Conley and Vorhaus, *supra* note 252.

[256] *Mayo Collaborative Services v. Prometheus Labs., Inc.* 130 S. Ct. 3543 (2010).

[257] *Prometheus Labs. Inc. v. Mayo Collaborative Services* 628 F.3d 1347 (Fed. Cir. 2010).

[258] 566 U.S. _(2012) (Slip opinion at 8–11).

[259] *Id.*, at 16.

[260] *Association for Molecular Pathology, et al. v. Myriad Genetics, Inc., et al.* 569 U. S. ____ (2013) (Docket no. 12–398)

[261] *Id.*, at 10–18.

[262] *Id.*, at 16–17.

## 2.2 Patentability of Biotechnology in the USA

The court, however, maintained that the case 'does not involve method claims, patents on new applications of knowledge about the BRCA1 and BRCA2 genes, or the patentability of DNA in which the order of the naturally occurring nucleotides has been altered'.[263]

Highlighting the scientific complexity involved in the case, Justice Scalia gave his concurring opinion:

> I join the judgment of the Court, and all of its opinion except Part I–A and some portions of the rest of the opinion going into fine details of molecular biology. I am un-able to affirm those details on my own knowledge or even my own belief. It suffices for me to affirm, having studied the opinions below and the expert briefs presented here, that the portion of DNA isolated from its natural state sought to be patented is identical to that portion of the DNA in its natural state; and that complementary DNA (cDNA) is a synthetic creation not normally present in nature.[264]

The logic given by the Supreme Court regarding patenting of cDNA has invited some criticism from different corners. Arti Rai and Robert Cook Deegan from Duke University, who work extensively in the field of biomedicine, human genetics and intellectual property, have criticized the reasoning of the court: 'Because intron removal is relatively routine, the Court's decision could be seen as stepping back to some degree from the Court's unanimous but highly controversial opinion last year in Mayo *versus* Prometheus,'[265] where the court maintained that 'adding scientifically routine activity to a "law of nature" is insufficient for patent eligibility'.[266] They further added that '[t]he Court's analysis does not connect the dots as to why claims to information in the form of cDNA are less problematic than claims to information in the form of gDNA'.[267]

Anand Mohan Chakrabarty, whose genetically modified *Pseudomonas* bacteria became the first genetically-engineered organism to gain a US patent, as a result of the Supreme Court decision in *Diamond v. Chakrabarty*, has also been critical about the Supreme Court's decision in *Association for Molecular Pathology versus Myriad Genetics*. He observes:

> Since thousands of isolated and purified genes from various sources have been patented, it is hard to revoke all such patents by simply saying that such procedures do not involve any inventive steps. Much efforts over the years were spent just to localize the two genes that were believed to be tumor suppressor genes and where specific mutations led to a loss of this tumor suppressing activity of breast and ovarian cancer. A much better scientific

---

[263] *Id.*, at 17–18.

[264] *Association for Molecular Pathology, et al. v. Myriad Genetics, Inc., et al.* 569 U. S. ____ (2013) (SCALIA, J., concurring at 1).

[265] Arti K. Rai & Robert Cook Deegan, "Moving beyond "Isolated" Gene Patents" 341 *Science* 137–138, available at http://www.sciencemag.org/content/341/6142/137 (lat visited on June 26, 2014).

[266] *Ibid.*

[267] *Ibid.*

rationale would have been to reject patent eligibility because of a lack of demonstrated utility of the isolated and purified BRCA1 and BRCA2 genes, as we have argued recently.[268]

The decision would have modest effect on human diagnostic testing field as many human gDNA patents have already expired or will expire soon.[269] The most obvious impact of the decision would be 'increased access, reduced price and perhaps most importantly, the emergence of multi-gene first-line genetic tests for inherited risk of breast and ovarian cancer—replacing the current multi-step process of testing first for just two genes'.[270]

### 2.2.2 Other Statutory Requirements

After passing the subject matter test, an invention has to fulfil other statutory requirements; novelty, utility, non-obviousness and written description.

**Novelty** Novelty marks a distinction between discovery and invention. For decades this distinction has manifested itself in the form of product of nature in the USA. For instance in the USA, 'isolated and purified compounds' are considered patentable subject matter, when the application meets the statutory criteria for patentability.

Sometimes the requirement of novelty overlaps with the subject matter requirement under Section 101 of the US Patent Act.[271] It is normally seen in the case of biotechnology inventions as most of them are related to naturally existing products and the determination of their novelty overlaps with the provision of exclusion within the product of nature category under the subject matter requirement. However, novelty moves a step forward from subject matter requirement—while the subject matter checks whether the product in question involves human intervention that would distinguish it from a naturally occurring material, the novelty requirement checks if the product which is not naturally occurring is new and different from the existing products.[272]

Sec. 102 of the US Patent Act provides a list of non-exhaustive conditions that negate the novelty.[273] Existing Patent Act contains provision regarding prior conception under Section 102(g).[274] However, this provision has been deleted by Leahy-Smith America Invents Act 2011, which introduces first inventor to file system

---

[268] Anand Mohan Chakrabarty, "Patenting human genes and mutations: A personal perspective" 19 Journal of Commercial Biotechnology" 2–5 (2013), available at http://commercialbiotechnology.com/index.php/jcb/article/view/626/576 (last visited June 26, 2014).

[269] Rai and Deegan, *supra* note 265.

[270] Ibid.

[271] 35 U.S.C. Sec. 101 prescribes, "Whoever invents or discovers any new and useful process, machine, manufacture, or composition of matter, or any new and useful improvement thereof, may obtain a patent therefore, subject to the conditions and requirements of this title."

[272] Kankanala (2007).

[273] 35 U.S.C. Sec.102.

[274] 35 U.S.C. Sec. 102(g).

## 2.2 Patentability of Biotechnology in the USA

in the USA to be applied to applications with effective filing dates after 16 March 2003.[275]

Sec. 102(g) states that a person shall be entitled to a patent unless 'before the applicant's invention thereof the invention was made in this country by another who had not abandoned, suppressed, or concealed it'.[276] The section relates to prior inventorship by another person in the USA and retains the rules governing the determination of priority of invention. The Section further provides that '[i]n determining priority of invention there shall be considered not only the respective dates of conception and reduction to practice of the invention, but also the reasonable diligence of one who was first to conceive and last to reduce to practice, from a time prior to conception by the other'.[277]

The Section makes it clear that in order to get the priority for patentability an inventor shall not only conceive the invention first but also be diligent in reducing it to practice. As per the Section, the inventor that conceives of the invention first and is diligent in reducing it to practice gets priority for patentability.[278] Here, conception connotes the formation of a definite and permanent idea of the complete and operative invention in the mind of the inventor.[279] Actual reduction to practice in context of a conceived invention can be deduced when the claimed invention work for its intended purpose.[280] However, a constructive reduction to practice is taken into account when a patent application on the claimed invention is filed in the patent office.[281] The date of the conception of an invention plays an important role in determining the prior conception.

In the context of biotechnological inventions or gene-based inventions, laboratory notebooks have gained importance in proving prior art conception and diligence in reducing the invention into practice. There is a bunch of judicial decision which has laid down different standards in determining conception, diligence and reduction to practice. *Hybritech versus Monoclonal Antibodies Inc.* points out that prior art which seeks to negate novelty of an invention, should contain all elements and limitations of the invention.[282]

In *Amgen Inc. versus Chugai Pharmaceutical Co. Ltd.*,[283] the court held that conception of a method of sequencing the gene would not be enough to satisfy conception of the gene sequence. Conception is said to be complete only if the gene can be

---

[275] Sec. 3 of the Leahy-Smith America Invents Act 2011 (enacted on September 16, 2011).
[276] 35 U.S.C. Sec. 102(g).
[277] *Ibid.*
[278] Kankanala, *supra* note 272, at 43.
[279] *Coleman v. Dines*, 754 F.2d 353, 359, 224 USPQ 857, 862 (Fed.Cir. 1985).
[280] *Great Northern Corp. V. Davis Core & Pad Co.*, 782 F.2d 159, 165, 228 USPQ 356, 358 (Fed. Cir. 1986).
[281] *Weil v. Fritz*, 572 F.2d 856, 865 n. 16, 196 USPQ600, 608 n. 16 (CCPA 1978).
[282] *Hybritech v. Monoclonal Antibodies Inc.*, 802 F.2d 159,165,228 USPQ356, 358 (Fed.Cir.1986).
[283] *Amgen*, *supra* note 163.

described by its physical properties or structure. In order to amount to conception, the idea should be workable by a person with ordinary skill in the art.[284]

The Court had made it clear in *Fiers versus Revel*[285] that it is not enough to define an invention solely by its principal biological property because an idea having no more specificity than its biological property would not be a conception. It has to be defined by its structure, name, formula, or definitive chemical or physical properties. Unlike in *Amgen*, the court in this case said that conception can be proved even without carrying out the invention as long as the conception is clear enough to enable a person with ordinary skill to carry it out.[286]

*Kridl versus McCormick*,[287] elucidates the utility of laboratory notebooks for proving conception. As per the court, conception of an invention should be clear enough to enable a person with ordinary skill in the art to carry it out and conception of utility of the invention can be inferred from the state of the art.[288] In *Brown versus Barbacid*[289] the Court maintained that conception must encompass all limitations of the claimed invention and was complete only when the idea was so clearly defined in the inventor's mind that only ordinary skill would be necessary to reduce the invention to practice, without extensive research or experimentation.[290] *Singh versus Brake*[291] points out that laboratory note books have to be corroborated by evidence to prove the date of conception.[292] *Invitrogen Corp. Clontech Laboratories, Inc.*[293] maintains that conception has to be corroborated by objective evidence and that expert testimony and arguments alone would not be sufficient.[294]

Section 102(a) contemplates that a patent will not be granted if the invention was known or used by others in the USA, or patented or described in a printed publication in the USA or a foreign country, before the invention thereof by the applicant for patent.[295] Section 102(b) provides that a patent will not be granted if the invention was patented or described in a printed publication in USA or a foreign country or in public use or on sale in the USA, more than one year prior to the date of the application for patent in the USA.[296]

In brief, as per Sections 102(a) and 102(b), an invention will not be novel, if it is known or used in USA, patented or published anywhere in the world and in public

---

[284] *Id.*, at 1206, 1207.

[285] *Fiers v. Revel*, 948 F.2d 1164 (C.A.Fed., 1993).

[286] *Id.*, at 1169, 1170.

[287] *Kridl v. McCormick*, 05 F.3d 1446 (C.A. Fed. 1997).

[288] Kankanala, *supra* note 272, at 49.

[289] *Brown v. Barbacid*, 276 F.3d 1327 (C.A.Fed., 2002).

[290] Kankanala, *supra* note 272, at 50.

[291] *Singh v. Brake*, 48 Fed.Appx.766 (C.A. Fed., 2002).

[292] Kankanala, *supra* note 272, at 51.

[293] *Invitrogen Corp. Clontech Laboratories, Inc.*, 429 F.3d 1052, 1068, 1069 (C.A. Fed. (Md.) 2005).

[294] Kankanala, *supra* note 272, at 53.

[295] 35 U.S.C. Sec. 102(a).

[296] 35 U.S.C. Sec. 102(b).

use or sale anywhere. Section 102(b) grants inventors 1 year grace period for public disclosure before grant of patents.[297] To negate novelty of an invention, a prior patent or publication should contain all elements and limitations of the invention within a single reference.[298] As regards to gene patents, novelty of gene-related inventions has been challenged in the US courts based on prior patent or publication and public use and sale bar. *In re Crish*[299] points out that existence of a larger sequence and a method of sequencing in a prior publication would anticipate a fragment or portion of the whole gene sequence.[300] *In re Ngai*[301] guides that attaching instruction sheet to an invention would not be sufficient to satisfy the novelty requirement in order to make it eligible for a patent.[302]

An invention will not be eligible for a patent if it forms part of public use within the USA for more than one year before the date of application.[303] The public use must be in natural and intended manner. The use need not be publicly accessible use, and secret commercial use would be considered to be public use. Public use of an invention for purposes of testing or experiment would not be considered to be public use.[304]

Further, an invention would not be novel if it is on sale in the USA for more than a year before priority date of the invention.[305] The Supreme Court has established two conditions for an on-sale bar: (1) the invention must be the subject of a commercial offer for sale, and (2) the invention must be ready for patenting. Secret commercial sale would give rise to an on-sale bar.[306]

**Novelty Requirement Under Sec. 102 as Amended by the Leahy-Smith American Invents Act 2011- (Applies to Applications with Effective Filing Dates after 16 March 2013.)** The novelty requirement under Section 102 has been simplified under Leahy-Smith America Invents Act, 2011, which provides that '[a] person shall be entitled to a patent unless the claimed invention was patented, described in a printed publication, or in public use, on sale, or otherwise available to public before the effective filing date of the claimed invention'.[307] One of the significant changes brought by the Leahy-Smith America Invents Act, 2011 is that novelty will be determined as of the effective filing date and not the date of invention.

---

[297] Kankanala, *supra* note 272, at 54.

[298] *Ibid*.

[299] *In re Crish*, 393 F.3d 1253 (C.A. Fed., 2004).

[300] Kankanala, *supra* note 272, at 55.

[301] *In re Ngai*, 91 Fed.Appx. 153 (C.A.Fed., 2004).

[302] Kankanala, *supra* note 272, at 57.

[303] 35 U.S.C. Sec. 102(b).

[304] Kankanala, *supra* note 272, at 58.

[305] 35 U.S.C. Sec. 102(b).

[306] Kankanala, *supra* note 272, at 58, citing *Pfaff v. Wells Elecs., Inc.*, 525 U.S. 55, 67, 119 S.Ct. 304, 142 L.Ed.2d 261 (1998).

[307] 35 U.S.C. Sec. 102 (Conditions for Patentability; Novelty) as amended by Sec. 3 of the Leahy-Smith America Invents Act, 2011.

Here, effective filing date is the actual filing date or the filing date of the earliest application to which the patentee or the applicant is entitled to claim priority.[308]

Apart from abolishing Sec. 102(g) as a source of prior art, the Leahy-Smith also abolished 'known or used by others' under Sec. 102(a) of the existing Patent Act. 'Public use' and 'on sale' are now prior art regardless of where in the world they occur. Further, there is no 1-year grace period as to third-party prior art unless that art was derived from the patentee, or unless the patentee disclosed first. AIA has added 'otherwise available to the public' as a source of prior art.[309]

Here, 'otherwise available to public' is somewhat ambiguous and open to wide interpretation. Now more conduct will be prior art, both because of the worldwide scope of sales and public uses and because there is no third party grace period. But the elimination of 102(g) (2) prior art will make anticipation somewhat less likely. Inventors are encouraged either to file or to disclose early, since disclosure will give them a 1-year grace period even against third party art. This grace period can be based not only on the inventor's actual disclosure, but also the disclosure of another subject to the same ownership or to a joint research agreement. This is similar to the current operation of Section 103(c).[310]

Under the Leahy-Smith America Invents Act, 2011, in priority disputes, priority is given to the first inventor to file an application or to the first inventor to disclose the invention to the public, assuming they file within a year. Swearing behind a reference is eliminated.[311] The first inventor to file system will prompt a race to patent office for patent filing. Here, there is some disagreement over the meaning of the term 'disclosure' that creates 1-year grace period. It is uncertain whether the term 'disclosure' would apply to any form of disclosure or cover only published disclosures. Here, it is notable that the grace period applies only to patentee's disclosures, and those made by people who got the invention from the patentee and not to third party.[312]

Laboratory notebooks, which are playing a significant role in proving the prior conception in biotechnology field, will lose its significance regarding prior conception under the Leahy- Smith America Invents Act, 2011, which replaces first to invent system by fist inventor to file system. In biotechnology field, which is full of uncertainty, first inventor to file system may lead to patenting of premature inventions.

**Utility** Sec. 101 of US Patent Act mandates utility requirement to be fulfilled before a patent is granted. It defines patent eligible subject matter as 'any new and useful process, machine, manufacture and composition of matter'.[313]

---

[308] 35 U.S.C. Sec. 100 as amended by Sec. 3 of the Leahy-Smith America Invents Act, 2011.
[309] Lemley (2011)
[310] *Ibid.*
[311] *Ibid.*
[312] *Ibid.*
[313] 35 U.S.C. Sec. 101.

## 2.2 Patentability of Biotechnology in the USA

In the USA, the Supreme Court defined the utility criterion for chemicals in *Brenner versus Manson*[314] case, Manson's patent application related to a chemical compound which was an intermediate for a steroid compound carrying anti-tumour properties. The patent applicant contended that the chemical compound was useful because an article revealed that an adjacent homologue of the steroid compound had been demonstrated to have tumour-inhibiting effects in mice.[315] He further added that the compound was useful because of its potential usefulness was under investigation by serious scientific researchers.[316]

The Supreme Court held that the fact that a homologue of the steroid compound had tumour-inhibiting effect in the mice would not make it patentable because the field was unpredictable and filled with ambiguities.[317] The court also mentioned that granting a patent on compounds such as in the present case having no specific and substantial utility in currently available form would not promote research on compounds and might even block research due to patent exclusivity.[318] The court made it clear that an invention should have specific and substantial utility in currently available form and further maintained that inventions with unknown functions such as gene sequences, which are subject of research, cannot be considered to satisfy the utility requirement.[319] Justifying its strict approach regarding utility requirement, the court emphasised that 'a patent is not a hunting license. It is not a reward for the search, but compensation for its successful conclusion'.[320]

This approach, however, has proven difficult to apply in a predictable fashion as technology advances.[321] Biotech community in the USA found that the approach adopted in *Brenner versus Manson* was unduly strict in light of technological advances. Accordingly, the CAFC made the utility test flexible in the case, *In re Brana*.[322] In this case, Brana's patent application was directed to anti-tumour compounds that exhibited better action and better action spectrum than known compounds of the same family (benzo(de)isoquiniolines). The patent office rejected the application on the ground that it showed no practical utility.[323] The USPTO maintained that the applicant failed to show use of the compounds for a particular disease and that the tests cited in the application did not show anti-tumour activity of the compounds.[324]

Rejecting the finding of patent office, the CAFC observed that the patent application disclosed that the compounds had better action and better action spectrum as anti-tumour compounds. The court maintained that the cell lines cited by the

---

[314] *Brenner v Manson*, 383 U.S. 519 (1966).
[315] *Id.*, at 533.
[316] *Ibid.*
[317] *Id.*, at 533.
[318] *Id.*, at 534.
[319] Kankanala, *supra* note 272, at 30.
[320] Brenner, *supra* note 314, at 536.
[321] Merrill and Mazza, *supra* note 17, at 81.
[322] *In re Brana*, 51 F.3d 1560 (1995).
[323] Lakshmikumaran (2007).
[324] *Brana*, *supra* note 314, at 1562.

applicants on which the compounds had been tested and proved effective were originally derived from lymphocytic leukaemia in mice and could, therefore, be considered to be useful against specific diseases.[325] The court further mentioned that there was enough evidence in the prior art in the form of utility of structurally similar compounds to prove usefulness of the compounds in question.[326] It was contended that the prior art has not been shown successful in clinical trials, but only in animal models. Replying to this contention, the court maintained: 'FDA approval, however, is not a prerequisite for finding a compound useful within the meaning of the patent laws... Usefulness in patent law, and in particular in pharmaceutical inventions, necessarily includes the expectation of further research and development'.[327] Eventually, the CAFC reversed the decision of the Board and granted the patent.

The court instructed the USPTO that an applicant's assertion of utility is presumptively correct unless based on implausible scientific principles. The court also emphasised that the burden is initially on USPTO to provide evidence showing that someone of ordinary skill in the art would reasonably doubt the asserted utility before it enters a rejection for lack of utility.[328]

USPTO issued revised utility guidelines[329] in 1995 after conducting public hearings on utility requirement in light of the CAFC's call for examiners to be more cautious while making utility rejections consistent with statutory standard.[330] Soon after 1995 guidelines were issued, it was realised that examination had become too lax. Consequently, in 2001 USPTO issued revised guidelines, raising the utility standard once again to limit the patenting of genes of unknown function.[331] The 2001 utility guidelines directed examiners to apply Supreme Court precedent and to inquire that the application asserts a 'specific and substantial utility that is credible'.[332] Here, specific utility indicates a utility which is particular to the subject matter claimed (e.g. a claim to a polynucleotide, which is disclosed as being useful as a gene probe or a chromosome marker, would not be specific if the target is not disclosed). A substantial utility denotes the real world use. Credibility of utility is established when a person skilled in the art acknowledges that the invention is currently available for the use defined.[333]

Although USPTO has adopted a more robust standard regarding utility criterion by issuing 2001 utility guidelines, however, these guidelines do not have the force of law without judicial endorsement.[334] A 2005 case *In re Fisher*[335] raised question

---

[325] *Id.*, at 1565.

[326] *Id.*, at 1566.

[327] *Id.*, at 1568.

[328] Merrill and Mazza, *supra* note 17, at 81.

[329] U.S. *Utility Examination Guidelines*, 60 Fed. Reg. 36,263 (July 14, 1995).

[330] Merrill and Mazza, *supra* note 17, at 82.

[331] USPTO, Utility Examination Guidelines, 66 Fed. Reg. 1092 (January 5, 2001).

[332] *Id.*, at 1098, quoted in Merrill and Mazza, *supra* note 17, at 82.

[333] Lakshmikumaran, *supra* note 323, at 51.

[334] Merrill & Mazza, *supra* note 17, at 82.

[335] *In re Fisher*, 421 F.3d 1365 (C.A. Fed., 2005).

regarding the utility standard in the context of claims to nucleic acid molecules. In the said case, the examiner rejected claims to ESTs encoding fragments of maize proteins for failure to disclose a specific and substantial utility for the claimed molecules. The USPTO Board of Patent Appeals and Interferences affirmed, rejecting as inadequate the asserted utilities of the ESTs for identification and detection of polymorphisms and for use as probes or as a source of primers. On appeal, the applicant's assignee (Monsanto) has contended that USPTO is applying a heightened utility standard to ESTs corresponding to genes of unknown function without statutory authority. Rejecting this contention, the CAFC upheld the Board's rejection and approved USPTO's guidelines, complying with the court's interpretation of the utility requirement.[336]

Concerns have been made that new advances in structural genomics and proteomics will pose great challenge before courts to resolve the utility issue in future. With the current stringent utility standard, 'whenever understanding of the function and uses of structures lags behind the discovery of the structures themselves, the determination of how much information on practical utility is necessary presents a line drawing problem'.[337] In this regard 'the successful interaction between USPTO and the scientific community about where to draw the line for ESTs provides a model for future interactions concerning how to apply patent law to new types of discoveries as science moves forward'.[338]

**Non-obviousness** The 1952 Patent Act was the first Patent Statute to impose an explicit requirement of non-obviousness. Section 103 of the US Patent Act provides that an applicant is not entitled to a patent 'if the differences between the subject matter sought to be patented and the prior art are such that the subject matter as a whole would have been obvious at the time the invention was made to a person having ordinary skill in the art to which the subject matter pertains'.[339] Non-obviousness in the field of biotechnology is a fact-intensive determination where potential success in experimentation and new properties of the invention carry significant weight.[340] In the USA, though the USPTO determines the non-obviousness of an invention, however, the courts guide the USPTO regarding such determination through their judicial decisions.[341]

The Supreme Court articulated certain factors to determine non-obviousness:

1. The scope and content of prior art;
2. The differences between prior art and the claims at issue;
3. The level of ordinary skill in the art at the time the invention was made; and
4. Objective evidence of non-obviousness, if any.[342]

---

[336] Merrill and Mazza, *supra* note 17, at 83.
[337] *Ibid.*
[338] *Ibid.*
[339] 35 U.S.C. Sec. 103.
[340] Lakshmikumaran, *supra* note 323, at 48.
[341] *Ibid.*
[342] *Graham v. John Deere Co.*, 383 U.S. 1, 17–18, 86 S.Ct. 684, 693–94 (1966).

Here, objective evidence of non-obviousness may include: commercial success; long felt but unsolved need; unexpected result; others' failure to solve the same problem etc.[343]

In *Hybritech Inc. versus Monoclonal Inc*,[344] the Court elucidates that existence of discrete prior art with missing links would not make an invention obvious. Emphasising the importance of prior art in determining obviousness of an invention, the court opined that objective evidence such as commercial success, failure of others, long felt need and unexpected results must be considered before a conclusion on non-obviousness is reached.[345]

*In re. O'Farrell*[346] clarifies that obviousness does not require absolute predictability of success but only a reasonable predictability of success and that, for many inventions, it is quite obvious that there is no absolute predictability of success until the invention is reduced to practice. *In re Vaeck*[347] elucidates that the existence of a prior art reference indicating expression of genes in one species does not make the expression of the gene in other species obvious.[348]

In an *en banc* decision,[349] the CAFC made it clear that if a patent application claims a molecule that is structurally similar to another useful molecule that is disclosed in the prior art, the claimed invention may be deemed prima facie obvious. Further, the burden shifts to the applicant to show that the claimed molecule has surprising or superior properties not possessed by the structurally similar prior art.[350]

In *Amgen Inc versus Chugai Pharmaceutical Co. Ltd.*,[351] Amgen's patent claim was directed to DNA sequences encoding erythropoietin (Epo). In a patent infringement suit filed by the appellants, the defendants claimed that the patent was obvious in the light of prior art and therefore invalid.[352] The defendants contended that as of September 1983, one of ordinary skill in the art would have had a reasonable expectation of success in screening a genomic DNA library by Lin's method in order to obtain Epo.[353] The court rejected the defendants' obviousness claim in the light of opinions expressed by the experts that it would have been difficult to isolate the Epo gene in the year 1983 based on prior art and that chances of isolating were 50 % because the probes being used were fully degenerate.[354] The court also rejected the defendant's obviousness claim based on homology with monkey gene by stating that though it might have been feasible perhaps obvious to try, to probe a human

---

[343] Lakshmikumaran, *supra* note 323, at 48.

[344] *Hybritech Inc. v. Monoclonal Inc.*, 802 F.2d 1367 (C.A. Fed. (Cal.), 1986).

[345] *Id.*, at 1380.

[346] *In re. O'Farrell*, 853 F.2d 894 (C.A. Fed., 1988).

[347] *In re Vaeck*, 947 F.2d 488 (C.A.Fed., 1991).

[348] Kankanala, *supra* note 272, at 76.

[349] *In re Dillon*, 919 F.2d 688 (Fed. Cir. 1990 (en banc).

[350] Merrill and Mazza, *supra* note 17, at 86.

[351] *Amgen*, *supra* note 163.

[352] *Ibid.*

[353] *Id.*, 1208.

[354] *Ibid.*

genomic DNA library successfully with a monkey cDNA probe, it did not indicate that the gene could have been identified and isolated with a reasonable likelihood of success because neither the DNA nucleotide sequence of the human EPO gene nor its exact degree of homology with the monkey Epo gene was known at the time.[355] Therefore, the court made a distinction between obvious to try and reasonable expectation of success and held latter as a significant factor for determining the obviousness of an invention.

Initially, the process of cloning a gene for a known protein required considerable creativity and skill, however, as the field advanced, it became an increasingly routine activity.[356] The processes of isolating human genes and determining the gene code for specific protein have become well known and documented.[357] In the light of recent genetic advances, scientists expressed their concern that the non-obviousness standard should exclude from patent protection the results of high throughput DNA sequencing that can be (and have been) performed by modestly competent research technicians in a mechanised discovery process.[358]

Considering this concern, patent examiners started rejecting patent applications claiming genes encoding proteins for which a partial amino acid sequence had previously been disclosed, based on the reasoning that 'when the sequence of a protein is placed into the public domain, the gene is also placed in the public domain because of the routine nature of cloning techniques'.[359] But this trend has been reversed by the CAFC in two cases, *In re Deuel*[360] and *In re Bell*.[361] In these cases, the CAFC rejected the arguments by the patent examiners that knowing a general method for identifying genes through the use of nucleotide probes, as well as the complete or the partial amino acid sequence of protein renders the DNA base sequences for that protein obvious.[362] The CAFC emphasised that while considering a patent claim to DNA sequences, the non-obviousness determination must focus on the DNA molecules as chemical compounds rather than on the method for isolating DNA.[363] CAFC maintains that 'any given DNA sequence (whether a full DNA sequence that encodes a gene or a mere sequence fragment) is obvious only if the prior art actually recites a similar or identical sequence and not simply a method for isolating the sequence'.[364] In other words, a DNA sequence can be non-obvious despite the fact that its isolation is easy and routine. Such a low standard for patentability requirements has paved the way for many biotech companies to obtain

---

[355] *Id.*, 1208.

[356] Merrill and Mazza, *supra* note 17, at 85.

[357] Nese, *supra* note 171, at 165.

[358] Merrill and Mazza, *supra* note 17, at 84.

[359] *Ex parte Deuel*, 33 U.S.P.Q. 2d 1445, 1447 (Bd. Pat. App. & Int'f. 1993) (citing the view of PTO examiners).

[360] *In re Deuel*, 51 F.3d 1552 (Fed. Cir. 1995).

[361] *In re Bell*, 991 F.2d 781 (Fed. Cir. 1993).

[362] Rai (1999).

[363] *Ibid.*

[364] *Id.*, at 833–34.

patents on DNA sequence fragments that they have been able to isolate quickly through routine automated methods.[365]

In the present age of genomics, DNA sequence information is accumulated in genomic databases and the informational content of a gene becomes more important than its physical attributes. It is possible that two structurally similar DNA sequences can perform significantly different functions. In the light of structural similarity test adopted by the CAFC, it is very likely that there would be more prima facie rejections for claimed DNA sequences that are structurally similar to previously disclosed DNA sequences. To overcome these rejections, an applicant 'must show some surprising properties that will be difficult to do through mere biology *in silico* without further laboratory research to characterise the sequence more fully and to distinguish it from prior art'.[366]

*Obviousness Under Leahy-Smith America Invents Act 2011* The definition of obviousness has been amended by the AIA. Under the AIA, an invention is considered obvious 'if the differences between the claimed invention and the prior art are such that the claimed invention as a whole would have been obvious before the effective filing date of the claimed invention to a person having ordinary skill in the art to which the claimed invention pertains'.[367] Obviousness is now tested as of the effective filing date of the patent and not at the date on which the invention is made. This could make findings of obviousness marginally more likely as it changes the level of knowledge of person having ordinary skill in the art (PHOSITA).[368]

**Written Description** The last patentability requirement for an invention is enshrined in 35 USC Section 112, which contains provisions regarding written description, enablement and best mode. Section 112 requires that 'the specification shall contain a written description of the invention, and of the manner and process of making and using it, in such full, clear, concise, and exact terms as to enable any person skilled in the art to which it pertains, or with which it is nearly connected, to make and use the same, and shall set forth the best mode contemplated by the inventor of carrying out the his invention'.[369] The specification must also contain claims 'particularly pointing out and distinctly claiming the subject matter which the applicant regards as his invention'.[370]

The written description requirement in paragraph one of Sec. 112 provides the technical and background explanation necessary for one to read and understand the patent application, including its claims.[371] The patentee is not required to describe

---

[365] *Id.*, at 834.

[366] Merrill and Mazza, *supra* note 17, at 86.

[367] Sec. 103. Conditions for patentability; non-obvious subject matter as amended by Sec. 3 of the Leahy-Smith America Invents Act, 2011 (Applies to applications with effective filing dates after March 16, 2013).

[368] Lemley, *supra* note 309.

[369] 35 U.S.C. Sec. 112.

[370] *Ibid.*

[371] Gulliford (2004).

the claimed subject matter exactly to fulfil the written description requirement; however, the description must 'clearly allow persons of ordinary skill in the art to recognize that [the inventor] invented what is claimed'.[372]

The enablement requirement under Sec. 112 requires that the inventor must set forth in the patent specification enough information to enable one skilled in the art to make and use the invention without 'undue experimentation'.[373] The enablement requirement necessitates that the patentee must show others how to make and use the invention, presumably so competitors may improve upon the claimed invention.[374] The enablement requirement is also used by courts to effectively narrow down a claim.[375] Broad claims must be supported by an equally broad enablement, and if they are not, the inventor has not taught how to 'make or use' the invention, and the non-enabled claims will not be allowed.[376]

Finally, the first paragraph of Sec. 112 requires that the specification 'set forth the best mode contemplated by the inventor of carrying out his invention'.[377] The underlying purpose behind the best mode requirement is to prevent inventors from obtaining patent protection while keeping secret the best way to make their invention.[378]

Initially, there had been some confusion among the courts regarding existence of a separate requirement of written description other than enablement requirement. However, the CAFC cleared this confusion by mentioning that Sec. 112 of the Patent Act requires a written description of the invention, which is separate and distinct from the enablement requirement.[379] Emphasising on the distinction between written description and enablement requirement, the CAFC held that it is not enough to provide an enabling disclosure of how to make a product that is not described in the specification.[380] Further, disclosure of an amino acid sequence for a protein and a strategy for cloning the corresponding gene might be enough to satisfy the enablement standard, but it is not enough to satisfy the written description requirement, as elaborated by the CAFC.[381]

In several cases involving claims to DNA sequences, the CAFC maintained that 'written description' standard, which serves to ensure that the inventor was 'in possession' of the invention as of the patent application filing date, requires disclosure of information about the structure of products covered by the claim, not just a

---

[372] *Id.*, at 721 (quoting *Vas-Cath Inc. v. Mahurkar*, 935 F.2d 1555, 1563 (Fed. Cir. 1991).
[373] 35 U.S.C. Sec. 112.
[374] Gulliford, *supra* note 371, at 721.
[375] *Ibid.*
[376] *Ibid.*
[377] 35 U.S.C. Sec. 112.
[378] Gulliford, *supra* note 371, at 721.
[379] *Vas-Cath Inc. v. Mahurkar, 935 F.2d 1555, 1563 (Fed.Cir. 1991).*
[380] Merrill and Mazza, supra note 17, at 88 [citing *Amgen Inc. v. Hoechst Marion Roussel Inc.* 314 F.3d 1313, 1330 (2003), *Regents of University of California v. Eli Lilly*, 119 F.3d 1559, 1568 (Fed. Cir. 1997); *Fiers v Revel*, 984 F.2d 1164, 1170–71 (Fed. Cir. 1993)].
[381] *Id.*, at 89.

description of their function.³⁸² An adequate written description of a DNA requires more than a mere statement that it is part of the invention and reference to a potential method for isolating it. The disclosure of the actual sequence itself is required to satisfy the requirement of written description.³⁸³ Further, a mere disclosure of structure etc. of few species in a genus would not be sufficient to support a claim of the entire genus unless substantial features of the genus and substantial common physical characteristics are described.³⁸⁴ The deposit of genetic sequences may constitute adequate written description.³⁸⁵

The foregoing discussion of case law reflects that initially, the courts have adopted a heightened standard of written description requirement due to unpredictability in the art. However, as the field matured, the courts have relaxed the requirements by considering the advancements in the art.³⁸⁶ This trend is reflected in recent cases, *Capon versus Eshar*³⁸⁷ and *Falko-Gunter Falkner versus Inglis*.³⁸⁸ In *Capon*, the court held that there was no per se rule requiring recitation in the specification of the nucleotide sequence of claimed DNA when the sequence is already known in the field. Further, the nucleotide sequence need not be mentioned in the specification if it is well known in the field.³⁸⁹ In *Falko*, the court laid down that examples need not be cited in the written description, if the knowledge in the field is advanced. The court also made it clear that actual reduction to practice is not required if the knowledge in the field makes the conception of the invention concrete.³⁹⁰

Despite a relatively relaxed approach of the court in the abovementioned cases, the written description requirement is still stringent regarding biotechnology inventions.³⁹¹ The USPTO passed the revised guidelines³⁹² for written description in 2001, which are largely in conformity with the case law. The guidelines explain how gene-based inventions such as genes, ESTs, antisense, ORFs etc. would be considered for purposes of written description.³⁹³

**Enablement** In the biotechnological context, a potential avenue for limiting gene patents may be found in Sec. 112's enablement requirement.³⁹⁴ The CAFC

---

³⁸² *Id.*, at 88 [citing *Regents of the University of California v. Eli Lilly* 119 F.3d 1559 at 1568 (Fed. Cir. 1997); *Fiers v. Revel.*, 984 F.2d 1164, 1171 (Fed. Cir. 1993).]

³⁸³ Kankanala, *supra* note 272, at 100 [citing *Fiers v. Revel.*, 984 F.2d 1164, 1170 (Fed. Cir. 1993)].

³⁸⁴ *Id.*, at 102 [citing *Regents of the University of California v. Eli Lilly & Co.* 119 F.3d 1559, 1569 (Fed. Cir. 1997)].

³⁸⁵ *Enzo Biochem Inc. v. Gen-Probe Inc., 323 F.3d 956 (C.A. Fed. (N.Y.) 2002).*

³⁸⁶ Kankanala, *supra* note 272, at 108.

³⁸⁷ *Capon v. Eshar (Capon)*, 418 F.3d 1349 (C.A.Fed., 2005).

³⁸⁸ *Falko-Gunter Falkner v Inglis*, 448 F.3d. 1357 (C.A.Fed., 2006).

³⁸⁹ Kankanala, *supra* note 272, at 106.

³⁹⁰ *Id.*, at 107.

³⁹¹ *Id.*, at 108.

³⁹² USPTO Guidelines and Training Materials on Written Description, Federal Register/Vol. 66, No. 4/Friday, 5 January, 2001.

³⁹³ Kankanala, *supra* note 272, at 107.

³⁹⁴ Nese, *supra* note 171, at 162.

invalidated a patent that claimed DNA sequence consisting of amino acid sequences duplicative of EPO in *Amgen*[395] case taking recourse of Sec. 112.[396] The court ruled that the generic claim directed to an isolated DNA sequence was not enabled when the claim read on about 4000 nucleotides but only disclosed how to make a few examples.[397] Since 'the scope of the enablement [must be] as broad as the scope of the claim', the overly broad claim at issue in *Amgen* violated the enablement requirement.[398]

*In re Vaeck*[399] involved a patent claim on the expression of endotoxin proteins in all strains of cyanobacteria, as opposed to any particular genus or species of cyanobacteria. The court rejected the claims to the extent that claims were too general to enable a person skilled in the art to make and use the claimed invention without undue experimentation.[400] *Enzo Biochem Inc versus Calgene Inc.*[401] elucidates that necessity of undue experimentation to carry out the claims makes the disclosure non-enabling.[402] Further, the level of a person skilled in an unpredictable art would be higher.[403] In *Falko-Gunter Falkner versus Inglis*,[404] the court laid down that if knowledge in one species can be easily applied to another species with the aid of the knowledge of a person with ordinary skill in the art, the description would be enabling.[405] So the courts have applied a heightened enablement standard for genetic inventions due to unpredictability in the field.[406]

Though enablement requirement limits the scope of a claim to a large extent, however, it has not been proved as effective as total ban on gene patenting.[407] For instance, under *Amgen*, as long as the patent description of the genes or methods it intends to cover reasonably correlates to the scope of the claims, the enablement requirement will be satisfied.[408] It is therefore possible that applicants for gene patents would easily pass the enablement test by listing the subject matter. In conclusion, the enablement requirement would minimally diminish the harm caused by gene patents and leave most diagnostic gene patents relatively untouched, as the claimed methods would likely be described in detail in the patent specification. In other words, a relatively broad claim in a diagnostic gene patent would not be considered

---

[395] *Amgen, supra* note 163.
[396] Nese, *supra* note 171, at 162.
[397] *Ibid.*
[398] *Ibid*, quoting Amgen, supra note 163, at 1212.
[399] *Vaeck, supra* note 347.
[400] *Id.*, at 494.
[401] *Enzo biochem Inc. v. Calgene Inc.*, 188 F.3d 1362 (C.A. Fed. (Del.), 1999).
[402] *Id.*, at 1373.
[403] *Id.*, at 1374.
[404] *Falko, supra* note 388.
[405] Kankanala, *supra* note 272, at 114.
[406] *Ibid.*
[407] Nese, *supra* note 171, at 163.
[408] *Amgen, supra* note 163, at 1214.

unduly broad in the light of numerous examples a patentee would likely describe in the specification.[409]

Best mode requirement has not been used much and the Leahy-Smith America Invents Act 2011 effectively eliminates best mode as a ground for invalidity, unenforceability or cancellation by the PTO. The Act, however, appears to preserve it as a ground for PTO examination.[410]

## 2.3 Patentability of Biotechnology in European Union

There exists a divergence in the patent approaches regarding biotechnology inventions among member states of the European Union despite various efforts of harmonisation and unification of patent laws.

### 2.3.1 Traces of a Unified System of Patents for European Union

*Convention on the Unification of Certain Points of Substantive Law for Invention* The traces of legislative response to biotechnology inventions go back to the early 1960s when 1963 Convention on the Unification of Certain Points of Substantive Law for Invention was adopted. This convention has two main objectives: first, to give industry a greater degree of certainty about whether it could secure protection for any given invention on a broad geographical basis; and second, to contribute to the creation of an international patent.[411] The adoption of the said convention was a successful event as the European Convention 1973 and trade-related aspects of intellectual property rights (TRIPS) contain many provisions from it.[412]

*The European Patent Convention (EPC)* In 1973, at a diplomatic conference in Munich, the European Patent Convention (EPC) was adopted. The EPC allows a single patent application to be filed and examined with the European Patent Office (EPO). Based on the examination, the EPO grants or rejects a patent. Once a patent is granted through the EPO, it is nationalised in individual countries designated by the applicant. The enforcement of the said patent in each individual country is governed by their respective national laws. Here, it is likely that a patent is invalidated in one country but enforced in another country. This brings divergence and non-uniformity in patent practices among different member countries. However, the EU sets minimum standards for patent protection by issuing directives that member

---

[409] Nese, *supra* note 171, at 163.
[410] Lemley, *supra* note 309.
[411] *Dutfield, supra* note 17, at 202.
[412] *Id.*, at 202–203.

states must implement into their national law.[413] The EPO has an appeal system consisting of various boards,[414] the most important being the Enlarged Board of Appeal. While their judgements are not legally binding, national courts tend voluntarily to accept their authority.[415]

### 2.3.2 Specific Legislative Response to Biotechnology Inventions

European Union confronted with more organised opposition and controversy regarding the patenting of biotechnology as compared to the USA. However, gradually, European governments and the European Commission realised the enormous growth of biotechnology in the USA and Japan. Industries persuaded the governments and the Commission to provide strong patent protection to biotechnological inventions in order to remain in the race of biotechnology developments.[416]

*Directive on the Legal Protection of Biotechnological Inventions 1989* There had been a legal uncertainty regarding the biotechnology inventions till 1989 when the European Commission realised the fear that some European countries might respond to biotech patent controversy by simply banning patents on living organisms and genes. Responding to this situation, the European Commission drafted a Directive on the Legal Protection of Biotechnological Inventions in 1989 to harmonise patent law relating to biotechnology around high minimum standards, while preventing member states from 'back-sliding'.[417] While passing the Biotech Directive, EU had two goals: first, to clarify the protection allowed to biotechnological inventions; and second, to provide some sort of harmonisation between the different member states on the patentability of biotechnology.[418]

*Opposition Against the Biotechnology Directive from EU Member States* Though the directive was welcomed and backed by industries, it was opposed by a large group comprising several environmental, developmental, religious and anti-genetic

---

[413] Bryan (2009).

[414] The Boards of Appeal at EPO consist of Enlarged Board of Appeal, a Legal Board of Appeal and 26 technical boards of appeal. Among these, technical boards of appeal and the Legal Board of Appeal deal with appeals filed in relation to decisions reached by the first instance in the patent grant procedure, i.e. the Receiving Section and the Examining, Legal and Opposition Division. On the other hand, the Enlarged Board of Appeal deals with cases referred to it either by one of the technical boards of appeal or by the Legal Board of Appeal or by the President of the European Patent Office for a decision on an important point of law in order to secure uniform application of the law. Further, with the entry of the revised European Patent Convention in December 2007, the Enlarged Board of Appeal also reviews the decisions of the boards of appeal upon the petition of a party; available at http://www.epo.org/about-us/boards-of-appeal/faq-boards-of-appeal.html (Visited on July 20, 2011).

[415] *Dutfield, supra* note 17, at 204–205.

[416] *Id.*, at 202.

[417] *Id.*, at 206.

[418] Jameson (2009) cited in Bryan, supra note 413, at 57.

engineering non-governmental organisations and of course some of the European governments.[419] In October 1998, an action was brought by the Dutch Government against the European Parliament and the Council of the European Union before the court of Justice of the European Communities to have the directive nullified on the grounds that the Directive has an incorrect legal basis and it breaches obligations under international law including the Convention on Biological Diversity and violates human rights. Subsequently, Italy and Norway supported the action by the Dutch Government. The court, however, dismissed the action in June 2001. Despite this dismissal, the opposition is still continued and few governments are not enthusiastic about the Directive.[420]

*Guidelines for the Examination in the European Patent Office, 2001 (EPO Guidelines)* In 2001 EPO has issued guidelines regarding the patent application examination, elaborating the key words.

## *2.3.3 Sources Governing Patent Grants in Europe*

The three main sources of law that govern patent grants in Europe are: the agreements of the EPC, Directive 98/44/EC of the European Parliament and the Council of the European Union on the Legal Protection of Biotechnological Inventions (Biotech Directive), and the national laws of the individual European states.[421] The issuance of a European patent immediately creates a collection of national patents, the enforceability of which is governed by the independent laws of the numerous contracting states.[422]

## *2.3.4 Biotechnology as a Patentable Subject Matter in European Union*

### 2.3.4.1 Patentable Subject Matter

The provision regarding patentable subject matter is set forth under Article 52 of EPC which provides, 'European patents shall be granted for any inventions which are susceptible of industrial application, which are new and which involve an inventive step'.[423] The patentability requirements, defined in EPC under Articles 57

---

[419] Dutfield, *supra* note 17, at 206–207.
[420] *Id.*, at 207.
[421] Zekos (2006).
[422] *Ibid.*
[423] Article 52, European Patent Convention.

## 2.3 Patentability of Biotechnology in European Union

(Industrial Application), 54 (Novelty) and 56 (Inventive Step), are analogous to the utility, novelty and non-obviousness requirement to obtain a patent in the USA.[424]

According to the EPC, any invention is eligible for patentable subject matter unless it falls within the list of excluded inventions provided in it.[425] The actual scope of patentability of biotechnology was clarified through the introduction of Biotechnology Directive. Further, in order to bring more clarity, the Administrative Council of the European Patent Organisation amended rule 23 of the Implementing Regulations of the EPC on 16 June 1999, and brought it in line with the Biotechnology Directive.[426]

Rule 23 (b) enables that for European patent applications and patents concerning biotechnological inventions, relevant provisions of the EPC will be applied and interpreted in accordance with the provisions of Chapter VI of the implementing regulations. It further added that the European Biotechnology Directive on the Legal Protection of Biotechnological Inventions will be used as a supplementary means of interpretation.[427]

**Unicellular Organisms—Microorganisms** Article 53(b) of the EPC excludes plant and animal varieties and essential biological processes for the production of plants and animals from the scope of patent protection but allows patenting on microbiological products and processes.[428]

Rule 23(b) (6) of the Implementing Regulations defines a 'microbiological process' as any process involving or performed upon or resulting in microbiological material. Rule 23 c(c) provides that a microbiological or other technical process, or a product obtained by such process other than a plant or animal variety is patentable subject matter.[429] In T356/93, microorganisms were held patentable as products of microbiological processes and microorganisms were defined as generally unicellular organisms with dimensions beneath the limits of vision, which can be propagated and manipulated in a laboratory.[430]

**Multi-cellular Organisms** *Plants*

Article 53 (b) of the EPC provides that European patents are not available for plant and animal varieties and essentially biological processes for the production of plants and animals.[431] Rule 23 c of the Implementing Regulations provides that plants and animals are patentable if the technical feasibility of the invention is not

---

[424] Zekos, *supra* note 421.

[425] *Supra* note 423.

[426] Rule 23 of the Implementing Regulations to the Convention on the Grant' of European Patents on 5 October, 1973 as last amended by decision of the Administrative Council of the European Patent Organisation of 9 December, 2004.

[427] Ibid.

[428] Art. 53(b), European Patent Convention.

[429] Rule 23b(6) and Rule 23c(c) of the Implementing Regulations, *supra* note 413.

[430] Kankanala, *supra* note 272, at 21.

[431] Art. 53(b), European Patent Convention.

confined to a particular plant or animal variety. Genetically modified animals and plants have been held to be patentable as they fall outside the scope of animal or plant variety.[432] Recital 30 of the Biotech Directive provides that a plant grouping which is characterised by a particular gene (and not its whole genome) is not covered by the protection of new varieties and is, therefore, not excluded from patentability even if it comprises new varieties of plants.[433] In a case[434] relating to patentability of transgenic plants into which DNA had been inserted using recombinant DNA technology, the Technical Board of Appeals held that genetically modified plants were patentable subject matter.[435]

*Animals* Patentability of genetically modified animals was contested in *Harvard Oncomouse* case. The case involved patenting of a genetically altered mouse in which an activated oncogene was inserted to develop cancer. In 1989, the Examination Division rejected the patent application on the basis that oncomouse falls within the scope of Art. 53 (b), that excludes animal varieties from the patentable subject matter.[436] On appeal, the EPO Technical Board of Appeal held that Article 53(b) did not exclude per se, the patenting of animals.[437] The Board maintained that the test to be applied under Article 53(a) was one of 'unacceptability', based on the weighing up of potentially detrimental effects of the grant of a patent on the one hand and the invention's usefulness to the humankind on the other.[438] Applying this test, the Examination Division concluded that potential medical benefits of the mouse outweighed any concerns of animal suffering and risk of escape into the environment and, therefore, did not violate Art. 53(a) of the EPC. Consequently, the Examining Division granted patent on Harvard Oncomouse (European Patent No. EPO169672).[439]

However, in 2003, the EPO Opposition Division considered the Harvard Oncomouse patent application afresh in the light of the Biotechnology Directive and affirmed that living matter, particularly plants and animals could be patented.[440] It further added that the exclusion regarding animals under Art. 53(b) was limited to animal varieties only and that could not be extended to animals in general.[441] Despite its wide interpretation, the EPO Division maintained that the patent must be limited to transgenic rodents containing an additional cancer gene, rather than 'transgenic non-human mammals'.[442] Subsequently, in 2004, the EPO Technical

---

[432] 23 c of the Implementing Regulations, *supra* note 413.
[433] Kankanala, *supra* note 272, at 22 (citing Recital 30 of the Biotech Directve).
[434] NOVARTIS/Transgenic Plant, Technical Board of Appeal 3.3.4 [1999] E.P.O.R. 123.
[435] Kankanala, *supra* note 272, at 22.
[436] Harvard/Onco-Mouse, 1989 O.J. EPO 451.
[437] Harvard/Onco-Mouse, 1990 O.J. EPO 476.
[438] Rimmer, *supra* note 42, at 90.
[439] *Ibid*.
[440] Harvard/Oncomouse [2003] OJEPO 473.
[441] Rimmer, *supra* note 42, at 90.
[442] *Id*., at 91.

## 2.3 Patentability of Biotechnology in European Union

Board of Appeal further reduced the scope of the patent claims from transgenic rodents to transgenic mice.[443] The Board regretted the long delays involved in the case and eventually the patent in respect of the Harvard Oncomouse expired in 2005.[444]

*Gene* The EPC does not contain specific provisions regarding patentability of genes; however, the Implementing Regulations that adopted Biotech Directive contains specific provisions pertaining to patentability of genes.[445] The Implementing Regulations provide that, 'biological material, which is isolated from its natural environment or produced by means of a technical process even if it is previously occurred in nature', is patentable.[446] Biological material has been defined as 'any material containing genetic information and capable of reproducing itself or being reproduced in a biological system'.[447] Here, isolated genes as biological material could be a patentable subject matter because it can be reproduced in a biological system such as cell or bacteria.[448]

Rule 23 e of the Implementing Regulations specifically mentions that gene sequences are patentable. Rule 23e (2) provides:

> An element isolated from the human body or otherwise produced by means of a technical process, including the sequence or partial sequence of a gene, may constitute a patentable invention, even if the structure of that element is identical to that of a natural element. [449]

It follows from the Implementing Regulations that partial and complete gene sequences are patentable under the EPC.[450] In *Howard Florey/Relaxin*, the Opposition Division of the EPO held that a DNA sequence coding for Human H2 Relaxin was a patentable subject matter.[451]

*Human Being* Rule 23d of the Implementing Regulations excludes certain biotechnological inventions dealing with human being from patentability. It provides:

> Under Article 53(a), European Patents shall not be granted in respect of biotechnological inventions, which, in particular, concerns,
>
> a. processes for cloning human beings;
> b. processes for modifying the germ line genetic identity of human beings; uses of human embryos for industrial or commercial purposes;

---

[443] Harvard/Oncomouse [2004] T 0315/03–3.3.8.
[444] Rimmer, *supra* note 42, at 91.
[445] Kankanala, *supra* note 272, at 23.
[446] Rule 23 c (a) of the Implementing Regulations, supra note 413.
[447] Rule 23 b (3) of the Implementing Regulations, supra note 413.
[448] Kankanala, *supra* note 272, at 23.
[449] Rule 23 e (2) of the Implementing Regulations, supra note 413.
[450] Kankanala, *supra* note 272, at 23.
[451] Howard Florey/Relaxin (Opposition by Franktion der Grunen Im, Europaischen Parliament; Lannoye), Opposition Division [1995] E.P.O.R. 541.

c. processes for modifying the genetic identity of animals which are likely to cause them suffering without any substantial medical benefit to man or animal, and also animals resulting from such processes. [452]

Further, Rule 23e of the Implementing Regulation provides that '[t]he human body, at various stages of its formation and development, and the simple discovery of its element is not patentable'.[453] However, elements isolated from the human body or otherwise produced by means of a technical process are patentable even if the structure of that element is identical to that of a natural element.[454]

## 2.3.5 Other Statutory Criteria for Patents

**Novelty** Article 52 (1) of the EPC mandates that an invention must be new in order to be patentable.[455] Article 54 mentions about the determination of the novelty of any invention. It states:

1. An invention shall be considered to be new if it does not form part of the state of the art.
2. The State of the art shall be held to comprise everything made available to the public by means of a written or oral description, by use or any other way, before the date of filing of the European patent application.[456]

In the context of gene patenting, where an application is made in relation to some process involving the gene the fact that the sequence is already isolated and purified may prevent a patent from being granted in relation to the process unless the second process itself is sufficiently new to meet the novelty criteria.[457]

In Decision G2/88,[458] the Enlarged Board of Appeal of the EPO upheld the novelty of a claim with regard to the use of a known compound for a second purpose to that in respect of which a patent protection had already been granted. It was contended that the second use was not novel in the sense that the use was inherent in the substance itself knowledge of which has been made available to the public.[459] The EPO clarified the situation by mentioning that under Art. 54 of the EPC, a hidden or secret use of a substance has not been made available to the public and so is not a ground for refusing a patent. If the product of a gene is capable of two different uses

---

[452] Rule 23 (d) of Implementing Regulations supra note 413.
[453] Rule 23e (1) of Implementing Regulations *supra* note 413.
[454] Rule 23e (2) of Implementing Regulations *supra* note 413.
[455] Art. 52(1), European Patent Convention.
[456] Art. 54, European Patent Convention.
[457] Brian Cain, *Legal Aspects of Genetic Technology* 127 (Sweet & Maxwell Ltd, London, 2003).
[458] G2/88 [1990] O.J. EPO 93.
[459] Cain, *supra* note 457, at 128.

2.3 Patentability of Biotechnology in European Union

in two technical processes, two patents protecting each of those processes could be granted.[460]

In *Howard Relaxin*,[461] the Opposition Board made it clear that isolation of a gene of a known protein for the first time through conventional methods would make the gene sequence novel. Natural existence of genes would not anticipate their isolation, as the isolated genes containing only the coding regions were different from their counterpart.[462] Amgen decision[463] suggests that existence of general methods would not anticipate a specific process for isolating gene sequence of a particular protein. The case also points out that existence of large sequences or gene banks would not anticipate specific sequences.[464] Decision T 223/92 elucidates that references citing general information without specific inputs relating to the invention based on which a person with ordinary skill cannot deduce the invention will not negate novelty of a claimed protein or DNA sequence.[465]

A prior art reference has to be clear, certain and specific and should contain all elements in the subject matter of the claimed invention in order to negate novelty of a genetic invention.[466] Even a very small difference such as a single amino acid in the sequence of a gene or protein can make it novel and patentable. Further, existence of fragment, as a part of a longer sequence does not anticipate the fragment.[467] The Unilever Decision (T 386/94) lays down that a prior art reference discussing a partial sequence coding for a particular protein cannot negate the novelty of the complete sequence coding for the same protein.[468]

**Inventive Step** According to Article 54 of the EPC, the other requirement for patentability of an invention is that the invention claimed involves an inventive step.[469] Article 56 defines 'inventive step' as:

> An invention shall be considered as involving an inventing step if having regard to the state of the art, it is not obvious to a person skilled in the art.[470]

---

[460] *Ibid.*

[461] Howard Florey Relaxin (Oppositions by Franktion der Grunen Im Europaischen Parliament; Lannoye), Opposition Division, 8 December 1994, (1995) E.P.O.R. 541.

[462] Kankanala, *supra* note 272, at 61–62.

[463] *Kiren Amgen/Erythropoietin v. (Oppositions by Genzyme; Elanex Pharmaceuticals; Merckle; Boehringer Mannheim; Behringwerke; AKZO Pharma, Technical Board of Appeal* 3.3.4, 21 November 1994, (1995)E.P.O.R. 629.

[464] Kankanala, *supra* note 272, at 63.

[465] *R. v. Genentech/HIF-Gamma*, T 223/92, Technical Board of Appeal 3.3.2, 20 July, 1993, (2003) E.P.O.R. 12, 106, 107.

[466] Biogen/Human Beta-interferon (Opposition by Schering, Technical Board of Appleal 3.3.4, 8 April, 1997 T 207/94 (1999) E.P.O.R. 451, 453, 461, 462.

[467] BIOGEN/Hepatitis B Virus (Opposition by Abbott;Takeda; Warcoin;Smithkline Beecham;Institute Pasteur; Intervention by Medeva) TechnicalBoard of Appeal 3.3.2, 16 June, 1994,(1999) E.P.O.R. 361, 377, 378.

[468] Unilever/Chymosin (Opposition by Celltech; Hansens Laboratorium), T/386/94,Technical Board of Appeal 3.3.4, 11 January, 1996, (1996), (1997) E.P.O.R. 184. 193, 194.

[469] Art. 54, European Patent Convention.

[470] Art. 56, European Patent Convention.

Here, the term, 'obvious' means that the subject matter follows plainly or logically from the prior art. The level of skill attributed to the hypothetical person reading the prior art is relatively high in the genetic technology field as compared to any other field. However, this skill usually changes as the technology becomes a routine of life.[471] Since genetic technology involves complexities, the determination of the state of art ascertaining the inventive step is problematic.[472]

To cope with this situation EPO has developed a so called 'problem/solution approach' to examine and assess an application whether it satisfies the inventive step criterion. This solution involves three steps:

1. An objective assessment of the technical result achieved, accompanied by an analysis what constitutes its closest prior art against which the assessment is to be made;
2. Determination of the technical problem, which is to be solved and an analysis of the features of the invention;
3. Analysis whether the technical result achieved would have been obvious to the person skilled in the art.[473]

After determining the problem of the state of art, the next issue is what is meant by the person skilled in the art. In this context, EPO guidelines provide:

> [t]he person skilled in the art should be presumed to be an ordinary practitioner aware of what was common general knowledge in the art at the relevant date. He should be presumed to have had access to everything in the "state of art", in particular the documents cited in the search report, and to have had at his disposal the normal means and capacity for the routine work and experimentation. If the problem prompts the person skilled in the art to seek its solution in another technical field, the specialist in the field is the person qualified to solve the problem'.[474]

In Decision T0455/91,[475] the Board of Appeals held that a skilled person working in one area of genetic engineering, e.g. expression of gene in yeast, would regard a means found possible in a neighbouring area of genetic engineering e.g. the expression of genes in bacteria, as being useable in his own area, if this transfer of technical knowledge appears to be easy and involve no obvious risks.[476] In the present case, patent protection was denied on the basis that application of the genetic engineering technique to a different organism did not on its own constitute a sufficiently inventive step for the purposes of conferring patent protection.[477]

In *Harvard Oncomouse* case, it was held that a Nobel Laureate is not to be deemed representative of the average person skilled in the art.[478] As regards to the

---

[471] Cain, *supra* note 457, at 130.
[472] *Ibid.*
[473] Triazole case, T939/92, OJ EPO 1996, 309.
[474] *Guidelines for Examination in the European Patent Office,* 71 Part C Chapter IV.
[475] T0455/91 (Genentech) [1995], O.J. EPO684.
[476] Cain, *supra* note 457, at 131.
[477] Ibid.
[478] T60/89 (Harvard) [1995] O.J. EPO 268.

inventive step, the decision of the House of Lords in *Biogen Inc versus Medeva plc*[479] is worth mentioning. The House of Lords held:

> [w]henever anything inventive is done for the first time it is the result of the addition of a new idea to the existing stock of knowledge. Sometimes, it is the idea of using established techniques to do something, which no one had previously thought of doing. In that case the inventive idea will be doing the new thing. Sometimes it is finding a way of doing something which people had wanted to do but could not think how. The inventive idea would be the way of achieving the goal. In yet other cases, many people may have a general idea of how they might achieve a goal but not know how to solve a particular problem, which stands in their way. If someone devises a way of solving the problem, his inventive step will be that solution, but not the goal itself or the general method of achieving it.[480]

In *Genentech Inc's Patent*,[481] the court considered what was the appropriate level of skill and knowledge to be given to the hypothetical person skilled in the prior art. The court was of the opinion that with regard to an invention in the field of genetic engineering, it was appropriate to attribute to that person a degree of skill and inventiveness normally found in research teams working in this difficult field.[482]

The T2/83[483] elucidated that the requirement of inventive step would be met if a practitioner found the particular solution when confronted with the technical problem, not whether he generally could solve it by chance. There must be thus a 'reasonable expectation of success' on the part of an ordinary practitioner. The *Biogen*[484] case maintained a distinction between the 'reasonable expectation of success' and the hope of succeeding by holding that ordinary practitioner must reasonably predict already from the beginning that he would be able to solve the technical problem, not that he merely hopes to solve it. The threshold of inventive step would not be met when the problem was straightforward, even if it required a considerable amount of work.[485]

The Trilateral Project specifies the application of the prong of inventiveness to the EST's. DNA fragments, which do not have any specific utility, are not inventive. Also cloned sequences are not inventive, as it is routine in the field of genetic research, unless they have a new technical effect or specific utility. Thereby, if the EST's can be used to diagnose a specified disease or they can enable identification of a new sequence with known specific utility, they are held to be inventive because the technical effect would be present.[486]

Contrasting the USA' approach of evaluating the obviousness of nucleic acid molecules based on structural similarity, Europe normally evaluates inventive step

---

[479] *Biogen Inc v Medeva plc* [1997] R.P.C.1.
[480] Ibid.
[481] [1989] R.P.C. 147.
[482] Cain, *supra* note 457, at 131.
[483] T2/83, OJ EPO 1984, 265.
[484] Biogen case, T296/93, OJ EPO 1995, 627
[485] Unilever case, T386/94, OJ EPO 1996, 658.
[486] Trilateral Project 24.1 at 2.1., cases E and D].

based on the obviousness of the methods of isolation of the nucleic acid molecules.[487]

**Industrial Applicability** Article 52 (1) prescribes that an invention must be susceptible to an industrial application in order to be patentable.[488] Art. 57 of the EPC mentions about when an invention is susceptible of industrial application. It states:

> An invention shall be considered as susceptible of industrial application if it can be made or used in any kind of industry including agriculture.[489]

Here, industry does not mean that the genetic technology needs to be such that only machines can carry it out. However, it requires that the subject matter of the application must be useful and of practical rather than merely aesthetic value.[490]

Art. 52(4) of the EPC excludes methods for treatment of the human or animal body by surgery or therapy and diagnostic methods practised on the human or animal body from invention despite the fact that these are susceptible to industrial application.[491]

The EPO Guidelines clarify that genetic technology that is not directed towards surgery, therapy and diagnosis in the sense used in 52 (4) is capable of being patented:

> It should be noted that Art. 52(4) excludes only methods of treatment by surgery or therapy and diagnostic methods. It follows that other methods of treatment of live human beings and animals (e.g. treatment of a sheep in order to promote growth, to improve the quality of mutton or to increase the yield of wool) or other methods of measuring or recording characteristics of the human or animal body are patentable provided that (as would probably be the case) such methods are of a technical, and not essentially biological character (see IV, 3.4) and provided that the methods are susceptible of industrial application.[492]

The EPO Guidelines further clarify by stating that:

> [A] process of treating a plant or animal to improve its properties or yield or promote or suppress its growth e.g. a method of pruning a tree, would not be essentially biological since although a biological process is involved, the essence of the invention is technical; the same could apply to method of treating a plant characterised by the application of a growth stimulating substance or radiation.[493]

Art. 52(5) states that the provisions of Arts. 52(1) and 52(4) shall not exclude the patentability of any substance and composition comprised in the state of the art or in use in surgery or therapy and diagnostic methods provided that its use for any such method is not comprised in the state of the art.[494] Du Pont[495] involved patenting of

---

[487] Zekos, *supra* note 421.
[488] Art. 52(1), European Patent Convention.
[489] Art. 57 European Patent Convention.
[490] Cain, *supra* note 457, at 134.
[491] Art. 52(4), European Patent Convention.
[492] Guidelines for Examination in the European Patent Office, 57 Part C Chapter IV
[493] *Id.*, at 54d.
[494] Cain, *supra* note 457, at 134.
[495] Du Pont/Appetite Suppressant [1987] E.P.O.R. 6 (1986).

## 2.3 Patentability of Biotechnology in European Union

a chemical product which had both cosmetic and therapeutic effect. The case elucidates that in order to satisfy the industrial applicability requirement, it would be sufficient to show that an invention is being used in enterprise having commercial objectives and the fact that an invention had therapeutic effect would not make the invention unpatentable if it has other uses.[496]

*Harvard Mouse Case* In *Harvard Oncomouse*,[497] one of the grounds of rejection of the patent application by the Examining Division was lack of industrial application. However, on appeal, the Technical Board of Appeals held that the claims relating to the transgenic mice had industrial applicability because they could be used as animal models in the medical or pharmaceutical industry for testing materials suspected of being carcinogenic as well as for testing materials for the ability to confer protection against the development of neoplasms.[498] The Board also made it clear that analysis of the requirement of morality and public order under Article 53 should be done independently and should not be combined with the requirement of the determination of industrial applicability.[499]

Unlike the USA, which requires specific substantial and credible utility, it would be enough if the invention can be made and used in an industry in Europe. For instance, a DNA sequence having only research use might not be an eligible candidate for patents in the USA, however, it might pass the industrial applicability test in Europe as long as such use in the industry is known and is not speculative.

**Written Description** After meeting the foregoing criteria a patent will not be granted unless sufficient details of the invention are disclosed in the application. Art. 83 of the EPC states:

'The European patent application must disclose the invention in a manner sufficiently clear, and complete for it to be carried out by a person skilled in the art'. In the context of genetic technology, Art. 83 becomes vital as an application seeking patents on gene will have to provide specific details of the substantive and credible use of any industrial application of the gene if the application is to succeed.[500]

There are guidelines laid down in the Implementing Regulations to the EPC in relation to how a gene should be described. The Implementing Regulations provide guidelines as to how a gene should be described. Rule 27a of the Regulations deals with the description that should appear in applications that relate to nucleotide or amino acid sequences:

1. If nucleotide or amino acid sequences are disclosed in European patent application, the description shall contain a sequence listing conforming to the rules laid down by the President of the European Office for the standardised representation of nucleotide and amino acid sequences.

---

[496] Kankanala, *supra* note 272, at 41.
[497] Harvard/Transgenic animal, Technical Board of Appeal 3.3.8 [2005] E.P.O.R. 31 (2006).
[498] *Id.* at 342.
[499] Ibid.
[500] Cain, *supra* note 457, at 137.

2. The President of the European Patent Office may require that in addition to the written application documents, a sequence listing in accordance with paragraph 1 be submitted on a data carrier prescribed by him accompanied by a statement that the information recorded on the data carrier is identical to the written sequencing listing.
3. If a sequence listing is filed or corrected after the date of filing, the applicant shall submit a statement that the sequence listing so filed or corrected does not include matter which goes beyond the content of the application as filed.
4. A sequence listing filed after the date of filing shall not form part of the description.[501]

Rule 28 of the Implementing Regulations sets out the details relating to the deposit of biological material in connection with any application.[502]

*Pioneer*[503] decision elucidates that a patent application would not be enabling if it requires undue experimentation to make and use the invention. It maintains that fortuitous or chance event for reproducibility, in the absence of evidence that their frequency was sufficiently high to guarantee success, would constitute an undue burden making a patent application non-enabling.[504]

In *R. versus Massachusetts Institute of Technology*,[505] the Technical Board of Appeal held that the assessment of the sufficiency of a disclosure of an invention in a particular case depends on certain general parameters such as the character of the technical field, the average amount of the effort necessary to put into practice a certain written disclosure in that technical field, the time when the disclosure was presented to the public and the corresponding general knowledge, and the amount of the technical details disclosed in a document. It also maintained that in the context of gene-based inventions, the requirement of trial and error to carry out the invention would make the disclosure non-enabling as the field is fraught with ambiguities.[506]

*Mycogen*[507] decision prescribed that the enabling disclosure had to be commensurate with the claims in the patent and if a person with ordinary skill in the art was not in a position to carry out all claims based on the disclosure, the disclosure would lack enablement.[508]

---

[501] Rule 27 (a) of Implementing Regulations *supra* note 428.

[502] Cain, *supra* note 457, at 137.

[503] Pioneer/Oilseed Brassica, Technical Board of Appeal 3.3.4, 5th March, 2004 [2004] E.P.O.R. 41.

[504] *Id*, at 421, 422.

[505] *R. v. Massachusetts Institute of Technology/Biopolymers*, Technical Board of Appeal 3.3.4, 21st January, 1998 [2003] E.P.O.R. 16.

[506] *Id*., at 142.

[507] Mycogen/Modifying Plant Cells (Opposition by Unilever, Centerns, Sandoz, Monsanto, and Max Planc Institute), Technical Board of Appeal 3.3.4, 8th May, 1996 [1998] E.P.O.R. 114.

[508] Kankanala, *supra* note 272, at 118.

## 2.3 Patentability of Biotechnology in European Union

The Technical Board of Appeal in *R. versus Genetech*[509] held that a patent application relating to a gene would be enabling even if the experimentation required was burdensome as long as undue experimentation was not required. It further laid down that deposit of biological materials was not compulsory as long as an application could be enabled based on written description.[510]

In *Biogen*[511] case, the Board laid down that every single variant in the claim did not need to be disclosed or claimed. Deposits could supplement the disclosure as starting materials. It further maintained that disclosure could rely on functional characteristics in case of genetic inventions.[512] The Technical Board of Appeal laid down in Genentech[513] case that 'the key test for biological processes of general applicability, so far as sufficiency was concerned, was whether or not the process as such was reproducible'.[514]

The above case study shows that in Europe, initially, the enablement requirement was relatively easy and relaxed. *Genentech* and *Biogen* reflect this trend. However, recent cases such as *Pioneer* and *R. versus Massachusetts* show that the enablement requirement has become stringent as the field matured.[515] Nevertheless, the enablement requirement in the USA is stricter than that in Europe. This is because European patent law assumes a well-developed and more certain field of genetic science as compared to the . Therefore, the USA requires more details to fulfil enablement requirement. Further, in the USA, a patent applicant has to satisfy enablement and written description requirement separately as compared to Europe where both the requirements are considered together.[516]

**Ordre Public and Morality** Unlike the USA, morality and *ordre* public has a legal foundation in Europe. Art. 53(a) of the EPC contains specific provision as to *ordre* public and morality which provides that 'European patents shall not be granted in respect of inventions, the publication or exploitation of which would be contrary to "ordre public" or morality, provided that the exploitation shall not be deemed to be so contrary merely because it is prohibited by law or regulation in some or all Contracting States'.[517] Art. 6.1 of the Biotechnology Directive contains almost similar language used in Art. 53(a) of the EPC. It provides that '[i]nventions shall be considered unpatentable where there commercial exploitation would be contrary

---

[509] *R. v. Genentech*/HIF-Gamma, Technical Board of Appeal 3.3.2, 20th July, 1993 [2003] E.P.O.R. 12.

[510] Kankanala, *supra* note 272, at 119.

[511] Biogen/Recombinant DNA (Oppositions by Hoffman la Roche; Upjohn; Boehringer Ingelheim Zentrale; Bender; Cetus; Hoechst; Boehringer Mannheim, Technical Board of Appeal 3.3.2, 16th March, 1989 [1990] E.P.O.R. 190.

[512] Kankanala, *supra* note 272, at 122.

[513] GenetechI/Polypeptide Expression, Technical Board of Appeal 3.3.2, 27th January, 1988 [1989] E.P.O.R. 1.

[514] *Ibid.*

[515] Kankanala, *supra* note 272, at 124.

[516] *Id.*, at 126.

[517] Art. 53, European Patent Convention.

to ordre public or morality; however, exploitation shall not be deemed to be so contrary merely because it is prohibited by law or regulation'.[518] Art. 6.2 of the Biotechnology Directive specifies certain types of claim that are considered unpatentable on the basis of *ordre* public and morality as contained in Art. 6.2 of the said Directive. The claims are:

a. processes for cloning human beings;
b. processes for modifying the germ line genetic identity of human beings;
c. uses of human embryos for industrial or commercial purposes;
d. processes for modifying the genetic identity of animals which are likely to cause them suffering without any substantial medical benefit to man or animal, and also animals resulting from such processes.[519]

Further, Recital 38 of the Biotechnology Directive enables that processes, the use of which offend against human dignity, such as processes to produce chimeras from germ cells or totipotent cells of human and animals, should be excluded from patentability.[520] Emphasising the importance of public order and morality in the field of biotechnology, Recital 39 states:

> Whereas ordre public and morality correspond in particular to ethical and moral principles recognised in a Member State, respect for which is particularly important in the field of biotechnology in view of the potential scope of inventions in this field and their inherent relationship to living matter; whereas such ethical or moral principles supplement the standard legal examinations under patent law regardless of the technical field of the invention.[521]

The issue of morality and public order has been often raised in biotechnology patent cases. However, *ordre* public/morality exemption is very rarely used to exclude an invention from the scope of protection.[522]

In *Harvard Oncomouse* case, the Examining Division considered the patentability of animals in the light of *ordre* public and morality provisions under Art. 53(a) of the EPC. The Examining Division remarked that the idea of patenting higher organisms attracted severe criticism from the public from ethical and economic reasons.[523] The Examining Division further maintained that the purpose of Art. 53(a) of the EPC is to exclude from protection inventions likely to induce riot or public disorder, or to lead to criminal or other generally offensive behaviour.[524] The Technical Board of Appeal held that the decision as to whether or not Art. 53(a) EPC would bar to patenting would seem to depend mainly on a careful weighing up of the suffering of animals and possible risk to environment on the one hand and the

---

[518] Art. 6.1 of Biotech Directive.
[519] Art. 6.2 of Biotech Directive.
[520] Recital 38 of Biotech Directive.
[521] Recital 39 of Biotech Directive.
[522] Cain, *supra* note 457, at 151.
[523] Harvard/Oncomouse, Examining Division, 14th July, 1989 [1990] E.P.O.R. 10.
[524] *Id.*, at 11.

invention's usefulness on the other.[525] In the light of this decision, the Examining Division held that the invention at hand was not immoral and against public order.[526]

In *Howard Relaxin* case, the patent applicant had developed a process for obtaining H2-relaxin and the DNA encoding it.[527] The invention was objected on the ground that it was not patentable under Art. 53(a) of the EPC. It was contended that as the patent required taking of the tissue from a pregnant woman, the act amounted to an immoral act, which was against humanity. It was because the act involved making use of female condition (pregnancy) in a technical process oriented towards profit. It was also contended that the patenting of human genes encoding H2-relaxin amounted to a form of modern slavery because it involved the dismemberment of women and their piecemeal sale to commercial enterprises throughout the world. Further, it was contended that patent human genes would amount to patent life, which was intrinsically immoral act.[528]

The Opposition Division maintained that in such cases, the fair test to be applied was to consider whether it was probable that the public in general would regard the invention so abhorrent that the grant of patent rights would be unconceivable.[529] Having said that, the Opposition explained that there was no reason to perceive the isolation of mRNA from the tissue of pregnant woman as immoral because it was taken by consent and such practice was accepted by the public.[530] The Opposition Division further stated that the exploitation of the invention did not involve dismemberment and piecemeal sale of woman because gene cloning could be done in unicellular organisms and a woman was only required initially for isolating the tissue for which consent was taken.[531]

Responding to the objection that patenting human gene would amount to patenting life, the Opposition Division noted that the objection was unfounded because DNA was not life, but a chemical substance, which carried genetic information and could be used as an intermediate in the production of proteins, which might be medically useful.[532] Further, patenting of a single gene had nothing to do with the patenting of life.[533] In the light of these reasoning, the Opposition Division held that patenting human genes such as that encoding H2-relaxin was not immoral and, therefore, could not be rejected on the basis of Art. 53 (a) of the EPC.[534]

---

[525] Harvard/Oncomouse, T19/90, Technical Board of Appeal 3.3.2, 3rd October, 1990 [1990] E.P.O.R. 501.

[526] *Ibid.*

[527] Horward Florey/Relaxin (Oppositions by Franklin der Grunen In Europaischen Parliament; Lannoye) Opposition Division, 8th December, 1994 [1995] E.P.O.R. 541.

[528] *Id.*, at 549.

[529] *Id.*, at 550.

[530] *Ibid.*

[531] *Id.*, at 551.

[532] *Ibid.*

[533] *Id.*, at 551

[534] *Ibid.*

In *Plant Genetic System*[535] case, Technical Board of Appeal maintained that plant biotechnology per se could not be regarded as being more contrary to morality than traditional selective breeding because both traditional breeders and molecular biologists were guided by the same motivation, namely to change the property of a plant by introducing novel genetic material into it in order to obtain a new, and possibly, improved plant.[536]

In Europe, *ordre* public and morality provision has been regularly applied by EPO to applications pertaining to gene-based inventions. An invention will be denied patent protection on the basis of *ordre* public and morality under Art. 53 (a) of the EPC, if evidence can be shown, that its exploitation is against public order or morality. In Europe, a patent will be granted to over genetically modified animals only when the benefit to the mankind outweighs the suffering caused to animals. However, the potential unknown danger to society or environment will not prevent patenting of an invention. Human and human related patents are considered immoral and, therefore, not patentable in Europe.[537] Since European Patent laws have specific statutory provisions regarding morality and public order, therefore, there is higher probability of rejecting patents on ethical and moral grounds in Europe than in the USA which lack such statutory provision.[538]

## 2.4 Patentability of Biotechnology Inventions in Canada

Canada has adopted a relatively restrictive approach regarding biotechnology patents as compared to the USA and Europe. Canada maintains a distinction between higher and lower life forms, permitting patents only to latter. However, this approach appears to be loosened in recent years. In Canada, two popular Supreme Court decisions have clarified the position of patenting higher life forms; the *Harvard College versus Canada (Commissioner of Patents)*[539] and *Monsanto Canada Inc. versus Schemieser*[540] decisions. The two cases involved patents covering animals and plants, respectively, that had been modified by artificial genetic manipulation. The *Monsanto* decision has provided at least a modicum of patent protection to 'higher life forms'.[541]

The approach of Canada regarding patenting of biotechnology inventions has been equally influenced by both, the USA and Europe. This is because Canada does not have clear guidelines governing patentability requirements in the field of

---

[535] Plant Genetic System/Glutamine Synthetase Inhibitors (Opposition by Greenpeace), Technical Board of Appeal 3.3.4, 21st February 1995, T/356/93, {1995] E.P.O.R. 357.

[536] *Id.*, at 372.

[537] Kankanala, *supra* note 272, at 137.

[538] *Id.*, at 138.

[539] Harvard College v. Canada (Commissioner of Patents), 2002 SCC 76, [2002] 4 S.C.R. 45.

[540] *Monsanto Canada Inc. v. Schmeiser*, 2004 SCC 34, [2004] 1 S.C.R. 902.

[541] Zahl (2004).

2.4 Patentability of Biotechnology Inventions in Canada 81

biotechnology as compared to the USA and Europe. Further, there is limited case law on biotechnology patents. Due to the aforementioned reasons, Canadian patent office is influenced by the patent practices of the USA and European patent offices. Canadian courts have often referred US case law and practice to guide the development of Canadian Patent Law.[542] Canada's restrictive approach towards patenting of higher life forms echoes the persistent opposition against Harvard Oncomouse patent in Europe.

In Canada, despite lack of clear guidelines and limited case law regarding patenting of biotechnological inventions, the issue of biotechnology patenting has been seriously debated and discussed in the domestic circle. This resulted in the establishment of the Canadian Biotechnology Advisory Committee (CBAC) in November 2001 to provide the government of Canada advice on crucial policy issues pertaining to biotechnology.[543]

At the outset of the discussion regarding patentability of biotechnology inventions in Canada, it becomes pertinent to throw light upon statutory requirement for patenting.

## 2.4.1 Statutory Framework for Patenting

### 2.4.1.1 Patentable Subject Matter

The definition of invention under the Canadian Patent Act is essentially the same as that found in the USA statute (except the word 'art', which is absent in the US Patent Act 1952).[544] Section 2 of the Canadian Patent Act defines 'invention' as 'any new and useful art, process, machine, manufacture or composition of matter, or any new and useful improvement in any art, process, machine, manufacture or composition of matter'.[545] The requirement, previously imposed by the courts, that an invention also be non-obvious to one of skilled in the art, is now part of the Act (Section 28.3) as a result of the 1996 amendments.[546] So, an invention is patentable when it is new, useful and non-obvious (i.e. the result of inventive ingenuity). However, the scope of subject matter is limited by the exclusions from patentability.

Sec. 27(8) of the Canadian Patent Act specifically excludes 'any mere scientific principle or abstract theorem' from patentability.[547] Canadian Patent Act, however, does not contain any comprehensive list of exclusions. Similar to the USA,

---

[542] Anita Nador, "The Patenting of Biotechnology in Canada", available at http://www.samedan-ltd.com/magazine/12/issue/44/article/1245.(visited on 29/07/2012).

[543] *Ibid.*

[544] 35 U.S.C. Sec. 101 enunciates patentable subject matter as "[w]hoever invents or discovers any new and useful process, machine, manufacture, or composition of matter, or any new and useful improvement thereof, may obtain a patent therefore, subject to the conditions and requirements of this title."

[545] R.S.C. 1985, c. P-4, s. 2.

[546] *Id.* s. 28.3.

[547] *Id.* s. 27.8

Canadian courts have recognised numerous limitations to the patentable subject matter by interpreting the definition of invention under Section 2 of Canadian Patent Act. In 1972, the Supreme Court of Canada in *Tennessee Eastman Co. et al. versus Commissioner of Patents*[548] disallowed claims for methods of medical treatment. Accordingly, methods of medical or surgical treatment of humans and animals, including methods of preventive medicine are not patentable in Canada. However, methods of treating animals to derive an economic benefit as well as methods of diagnosing a disease or medical condition in a human being are patentable.[549]

Further, section 2 of the Patent Act has been interpreted to exclude multicellular differentiated organisms (i.e. higher life forms like animals, plants and seeds),[550] methods which occur essentially according to the laws of nature without any significant human technical intervention.[551] Regarding computer programs, computer algorithms per se are not patentable in Canada[552]; however, software may be protected by Canadian patent law if it meets the traditional criteria for patentability (i.e. it must be new, non-obvious and useful).[553] For example, the new and non-obvious software is patentable in Canada only if the software directly provided a functional real world useful result (and not merely the calculation of a mere algorithm).[554]

There is no '*ordre* public' or morality clause in the Canadian Patent Act similar to Article 53(a) of the EPC.[555] The issue of morality had been raised in *Harvard College* case, where the Supreme Court unanimously held that the Commissioner of Patents has no discretion to refuse a patent on the basis of public policy considerations independent of any express provision in the Patent Act.[556]

**Patenting of Unicellular Organism** *Re Application of Abitibi Co.*[557] recognised the patenting of microorganisms in Canada. In this case, the question before the Patent Appeal Board was whether a yeast culture meant for digesting a pulp mill waste product was patentable. In order to make the naturally occurring yeast workable on the spent sulphite waste materials, there required human intervention i.e. the yeast had to be acclimatized and treated in order to function. The Board held that the yeast culture capable of digesting spent sulphite liquor was patentable. The Board noted

---

[548] *Tennessee Eastman Co. et al. v. Commissioner of Patents* (1972), 8 C.P.R. (2d) 202.

[549] *Manual of Patent Office Practice* [MPOP] Chapter 12.04.02 "Medical treatment".

[550] *Harvard College v. Canada (Commissioner of Patents)*, 2002 SCC 76, [2002] 4 S.C.R. 45; Monsanto Canada Inc. v. Schmeiser, 2004 SCC 34, [2004] 1 S.C.R. 902.

[551] *Pioneer Hi-Bred Ltd. V. Canada (Commissioner of Patents)* [1989] S.C.R. 1623 [(1989), 25 C.P.R. (3rd), 257 (S.C.C.)] at pp. 263–265.

[552] *Schlumberger Canada Ltd. v. Commissioner of Patents* (1981) 56 C.P.R. (2d) 204 (FCA).

[553] MPOP Chapter 16.03.02 "Patentability and Programming".

[554] *Ibid.*

[555] Art. 53(a) of the European Patent Convention stipulates that European patent shall not be granted in respect of "inventions the commercial exploitation of which would be contrary to "ordre public" or morality;.."

[556] *Harvard College v. Canada (Commissioner of Patents)*, [2002 SCC 76, 116–121 per Bastarache J., and 89–102 per Binnie J.].

[557] *Re Application of Abitibi Co.*, (1982), 62 C.P.R. (2d) 81 (P.A.B).

that though the yeasts were based on naturally occurring varieties, however, they had been treated to develop sulphur resistance and were thus significantly different to the naturally occurring varieties.[558]

Further, the Board was convinced that the yeast could be reproduced uniformly, but was sceptical as to whether such uniformity would hold in cases of higher life forms due to natural variations. With this decision; the Board explicitly extended the scope of patentable subject matter to unicellular living organisms.[559] The Board explained:

> Certainly this decision will extend to all microorganisms, yeasts, moulds, fungi, bacteria, actinomycetes, unicellular algae, cell lines, viruses or protozoa; in fact to all new life forms which are produced en masse as chemical compounds are prepared, and are formed in such large numbers that any measurable quantity will possess uniform properties and characteristics... Whether it reaches up to higher life forms—Plants... or animals—is more debatable...[560]

**Patenting of Higher Life Forms (Multicellular Organism)** *Pioneer Hi-Bred Ltd. versus Canada (Commissioner of Patents)*[561]
In the present case, applicants sought to obtain patent rights to a new soybean variety known as 'Soybean Variety 0877', which was developed by using the artificial cross-breeding of three known varieties.[562] The specification describes the subject of the application as:

> [A] plant line cultivated naturally but resulting from the artificial cross-breeding of three known varieties to produce a new variety combining the desirable characteristics of each one; the claims related to the plant, the pod and the seed. The invention is said to be unique of its kind and not to have existed previously in nature; among its chief characteristics are its high oil content, early maturity, stable high yields, resistance to seed shattering, and in particular, disease resistance to races 1 and 2 of Phytophtora megasperma var sojae as well as moderate resistance to another fungal pathogen, Sclerotinia sclerotiorium.[563]

The patent examiner at the Canadian Patent Office rejected the application in the light of the definition of invention under Sec. 2 of the Patent Act. The examiner concluded that the new plant variety described in the application was not covered under the definition of invention enshrined in Sec. 2 of the Patent Act.[564] He further added that it is the general practice of the patent office to consider 'subject matter for a process for producing a new genetic strain or variety of plant or animal, or the product thereof'[565] non-patentable.[566] An appeal was made to the Commissioner

---

[558] Perry (2008).

[559] *Id.*, at 72–73.

[560] (1982), 62 C.P.R. (2d) 81 (P.A.B) [*Re Abitibi*] cited in *Perry, supra* note 559, at 73.

[561] *Pioneer Hi-Bred Ltd. v. Canada (Commissioner of Patents)* (1989), 25 C.P.R. (3d) 257 (S.C.C.).

[562] *Pioneer*, [25 C.P.R. (3d) p. 259].

[563] *Ibid.*

[564] *Ibid.*

[565] Sec. 12.03.01(a) of the Patent Office Manual cited in *supra* note 41.

[566] *Supra* note 562.

of Patents for a review of the examiner's decision. The Commissioner of Patent referred the matter to a Patent Appeal Board constituted by him, where the Board in its report affirmed the examiner's decision. The Board also indicated that a limited interpretation should be given to the language embodied in the definition of the term 'invention' and accordingly appellant's invention did not qualify as a manufacture.[567] Finally, it stated that:

> [t]he Commissioner has not only the right but the duty to determine if an application is directed to patentable subject matter and if, according to his determination, it is not patentable then he is permitted to refuse to grant a patent.[568]

On appeal, the majority of the Federal Court of Appeal observed that Canadian patent legislation does not expressly exclude living organisms from patentability. Marceau J. questioned the contention of the Commissioner of Patents that he could establish limits other than those chosen by Parliament to decide whether an invention is patentable. However, the focus of the Federal Court of Appeal was to determine whether the subject matter of the application, the new soybean variety, could be regarded as an invention in the sense in which the legislator understood this word.[569] He answered this question as follows:

> Besides, speaking of the intention of Parliament, given that plant breeding was well established when the Act was passed, it seems to me that the inclusion of plants within the purview of the legislation would have led first to a definition of invention in which words such as "strain", "variety" or "hybrid" would have appeared, and second to the enactment of special provisions capable of better adapting the whole scheme to a subject matter, the essential characteristic of which is that it reproduces itself as a necessary result of its growth and maturity.[570]

Emphasizing on the requirements of Sec. 36 of the Act, Pratte J. suggested that the documentation submitted by Hi-Bred showed that a degree of 'luck' had enabled its new variety.[571] He observed:

> It follows that even a complete and accurate disclosure by the appellant of everything that the alleged inventor did to develop the new plant would not enable others to obtain the same results unless they, by chance, would benefit from the same good fortune.[572] [at para 9]

Pratte J. rejected the viewpoint of the Patent Appeal Board in *Re Application of Abitibi*,[573] which held that depositing a new microorganism in a culture collection to which the public had access was sufficient to satisfy the requirements of Sec. 36(1) of the Patent Act. Differentiating his point of view from that of the Patent Appeal Board in *Re Application of Abitibi*, he opined:

---

[567] *Ibid.*

[568] *Pioneer*, [25 C.P.R. (3d) p. 260].

[569] *Ibid.*

[570] *Pioneer Hi-Bred Ltd. v. Canada (Commissioner of Patents)* [1987] 3 F.C. 14 cited in *supra* note 550, at 260.

[571] *Pioneer, supra* note 556.

[572] *Pioneer Hi-Bred Ltd. v. Canada (Commissioner of Patents)* [1987] 3 F.C. 9, cited in *supra* note 562, at 260.

[573] *Re Application of Abitibi Co.,* (1982), 62 C.P.R. (2d) 81 (PAB).

## 2.4 Patentability of Biotechnology Inventions in Canada

The use of the seeds deposited by the appellant is, in a sense, the use of the invention itself. Subsection 36(1), as I read it, requires that the description be such that third persons, who do not have access to the invention or anything produced by it, be enabled to reproduce it. This opinion conflicts with the conclusion reached by the Patent Appeal Board in the Abitibi case where it was held that depositing a new micro-organism in a culture collection to which the public had access was sufficient to satisfy the requirements of subsection 36(1). That conclusion of the Board was, in my opinion, clearly wrong and based on what I consider to be an untenable interpretation of the decision of the House of Lords in American Cyanamid Company (Dann's) Patent.[574]

On appeal, the Supreme Court upheld the rejection of the patent application claiming a new variety of soybean and dismissed the appeal for lack of sufficient disclosure despite deposition of seed sample.

It is worth mentioning that at the time the *Pioneer* case was decided, the Patent Act did not recognize sample deposition as a viable way to meet the 'enabling description' requirement of patent law. Canada was not then a member of the Budapest Treaty, which required participating countries to accept samples deposited in recognized depositaries as fulfilment of the 'description' requirement for inventions in the field of biology. There has been no decision so far which clearly establishes that the non-enablement finding of the Supreme Court in *Pioneer-Hybrid* could now be overcome by a sample deposit under the Budapest Treaty. The question seems to be of little significance in Supreme Court's decisions in *Harvard College* and *Monsanto* involving patenting of higher life forms.[575]

The court explained that genetic engineering can occur in two ways; first includes cross-breeding through hybridization, whereby there is human intervention in the reproduction cycle but such human intervention does not alter the actual rules of reproduction, which continues to obey the laws of nature; and second requires a change in the genetic code affecting the entire hereditary material i.e. the human intervention occurring inside the gene itself.[576] The court concluded that the genetic engineering relating to *Pioneer*'s soybean application falls under the first category.[577] The court maintained:

> The intervention made by Hi-Bred does not in any way appear to alter the soybean reproductive process, which occurs in accordance with the laws of nature. Earlier decisions have never allowed such a method to be the basis for a patent. The courts have regarded creations following the laws of nature as being mere discoveries the existence of which man has simply uncovered without thereby being able to claim he has invented them. Hi-Bred is asking this court to reverse a position long defended in the case-law...[578]

The court, however, avoided a decision as to the patentability of Pioneer's hybrid seeds under Section 2 of the Canadian Patent Act, disposing of the appeal on the

---

[574] *Pioneer Hi-Bred Ltd. v. Canada (Commissioner of Patents)* [1987] 3 F.C. 9–10, cited in supra note 562, at 261.

[575] Zahl, *supra* note 542.

[576] *Pioneer, supra* note 562, at 263–264.

[577] *Id.*, at 264.

[578] *Pioneer, supra* note 562, at 264–265.

sole ground of lack of sufficient disclosure.[579] Citing the Supreme Court decision in *Pioneer Hi-Bred* Case, the Manual of Patent Office Practice maintains that the 'degree of technical intervention' embodied in the claimed process is an important consideration in biotechnology. Accordingly, a process which occurs essentially according to nature, with no significant technical intervention by man, is not patentable e.g. a process for producing a plant by traditional cross-breeding techniques is not patentable.[580]

**Harvard College versus Canada (Commissioner of Patents)[581] (Harvard College) Case** In the present case, the Supreme Court of Canada considered an appeal against the decision of the Federal Court of Appeal that the Harvard Oncomouse was patentable subject matter under Canadian law. The Harvard College filed its Canadian patent application (with the application number 484, 723), entitled 'Transgenic Animals' on 21 June 1985, claiming priority from a corresponding US patent application (serial number 623,774), which was filed on 22 June 1984.[582] The transgenic mammal (i.e. the Harvard Mouse or oncomouse) had been modified to include an activated oncogene sequence in its genome that pre-disposes the mammal to developing cancerous tumours. The said mammal had been found useful for cancer research and treatment of cancerous tumours.[583]

In its patent claim, the Harvard College sought to protect the process by which oncomice are produced and the end product of that process i.e. the founder mice and the offspring whose cells are affected by the oncogene.[584] Claims 1 to 12 of the application were directed to transgenic non-human mammals whereby Claim 1 stated:

> "A transgenic non-human mammal whose germ cells and somatic cells contain an activated oncogene sequence introduced into said mammal, or an ancestor of said mammal, at an embryonic stage." Claim 12 was limited to the transgenic mouse. Claim 13 through 26 in the application related to other aspects of the invention including method claims.[585]

The patent examiner at the Canadian Patent office rejected the 12 claims pertaining to the transgenic mammals as the products of invention on the basis that higher life forms fell outside the definition of 'invention' in the Patent Act and, therefore, were not patentable subject matter.[586]

Referring to the decision by the Patent Appeal Board in *Re Application of Abitibi*,[587] the examiner noted that in *Abitibi* the Patent Appeal Board stated that

---

[579] *Id.*, at 265.

[580] *Manual of Patent Office Practice* [MPOP] Chapter 17.02.02 "Processes to Produce Life Forms".

[581] *Harvard College v. Canada (Commissioner of Patents)*, 2002 SCC 76, [2002] 4 S.C.R. 45.

[582] Garland and Smordin (2003).

[583] *Ibid.*

[584] Leder, P. and T. Stewart, 'Transgenic animals' (1985) Canadian Patent No: CA 1341442 cited in Rimmer, *supra* note 42 at 91.

[585] Garland and Smordin, *supra note* 583, at 162–163.

[586] Garland and Smordin, *supra* note 583, at 163.

[587] *Abitibi*, *supra* note 558.

## 2.4 Patentability of Biotechnology Inventions in Canada

the question of whether higher life forms such as plants or animal were patentable was 'more debatable'.[588] Further, the patent examiner also relied on the decision of the Federal Court of Appeal in *Pioneer Hi-Bred Ltd versus Canada (Commissioner of Patents)*,[589] in which the majority of the Court of Appeal upheld the rejection of the claims but on the basis of insufficient disclosure, without deciding the subject-matter issue.[590]

On appeal, the Commissioner of Patents refused to grant a patent for the product claims in 1995 after considering the decisions by the Federal Court of Appeal and Supreme Court in *Pioneer* case. While referring to the decision by the Supreme Court in the *Pioneer case*, the Commissioner noted that in the said case, the Supreme Court had deliberately chosen not to decide whether a new plant variety was a patentable invention under Section 2 of the Patent Act. The Commissioner came to the conclusion that 'neither he nor the Examiner was bound by the decision of the Federal Court of Appeal's decision in that case'. However, he recognised that the majority decision of the Federal Court of Appeal in Pioneer could be of 'high persuasive influence'.[591]

Harvard College made an appeal against the decision of Commissioner of Patents to the Trial Division of the Federal Court, where the court dismissed the appeal on 21 April 1998 on the basis that the subject matter of claims 1 through 12 of the patent application did not fall within the definition of 'invention' under Sec. 2 of the Patent Act.[592] The Federal Court opined that a transgenic mammal is not truly reproducible because too much is left to chance (i.e. beyond human control), including the chromosomal location of the transgene and the degree of the transgene expression. Considering this fact, the court concluded that the transgenic mammal was not sufficiently reproducible to be a 'composition of matter' or an 'article of manufacture' under the current Patent Act.[593]

Harvard made an appeal against the said decision to the Canadian Federal Court of Appeal, where on 3 August 2000; the majority of the Appellate Court determined that the oncomouse was a composition of matter.[594] The majority also held that the requirements of human control and reproducibility are implicit in the statutory requirement that an invention be 'useful' and noted that the Patent Commissioner and the trial judge at the Federal Court had applied a far too broad control test, one that was not contemplated by the utility requirement for an invention.[595] The Federal Court of Appeal then sent the case back to the Commissioner of Patents with the direction to grant the patent on transgenic animal claims.[596] Finally, the case reached

---

[588] Garland and Smordin, *supra* note 583, at 163.
[589] *Pioneer*, [1987] 3 F.C. 8, 14 C.P.R. (3d) 491 (C.A.).
[590] Garland and Smordin, *supra* note 583, at 163–164.
[591] *Id.*, at 164.
[592] *Id.*, at 165.
[593] Rimmer, at 91.
[594] *Ibid*
[595] Garland and Smordin, *supra* note 583, at 167.
[596] Rimmer, at 91.

to the Supreme Court of Canada, when in the name of Commissioner of Patents, the Attorney General of Canada filed an application to seek appeal to the said court. On 14 June 2001, the Supreme Court granted the application for appeal.[597]

On 5 December 2002, the Supreme Court of Canada, in a 5–4 decision, overturned the decision of the Federal Court of Appeal and held that a genetically modified non-human mammal (i.e. Harvard Oncomouse) was not patentable under the current Canadian Patent Act. The court rejected the claims 1 through 12 relating to product as non-patentable under the Canadian Patent Act.[598] Bastatrache J. delivered the judgement for the majority, with Binnie J. writing the dissent. Here, the majority comprised judges who expressed legal and ethical concerns regarding the patenting of higher life forms in the absence of any explicit legislative direction from the Canadian Parliament. The minority, on the other hand, was composed of judges who believed that the patenting of biological inventions was necessary to encourage research and development in new technologies.[599]

Though the case involved various issues, however, the Supreme Court focused on two important issues; first, construction of the statutory definition of 'invention'; and second, the determination as to whether a higher life form such as the transgenic mammal of Harvard's application fell within the definition of 'manufacture' and 'composition of matter'.[600] Replying to various environmental, animal and social policy issues raised by various groups including Greenpeace, Sierra club, animal rights groups and religious groups, both the majority and the dissent stated that the sole question in dispute was a legal one, namely the construction of the term 'invention' in the Patent Act, and that the social policy issues were not relevant to the resolution of this question.[601]

The majority in the present case interpreted the term 'invention' under the Canadian Patent Act, in contrast to the wide interpretation given to the said term by US Supreme Court in *Diamond versus Chakrabarty* as 'anything under the sun that is made by man' was patentable. Bastarache J. observed:

> I cannot, however, agree with the suggestion that the definition is unlimited in the sense that it includes 'anything under the sun that is made by man'. In drafting the Patent Act, Parliament chose to adopt an exhaustive definition that limits invention to any 'art, process, machine, manufacture or composition of matter'. Parliament did not define 'invention' as anything new and useful made by man'. By choosing to define invention in this way, Parliament signalled a clear intention to include certain subject matter as patentable and to exclude other subject matter as being outside the confines of the Act. This should be kept in mind when determining whether the words 'manufacture' and 'composition of matter' include higher life forms.[602]

---

[597] *Ibid.*

[598] *Harvard College v. Canada (Commissioner of Patents)*, [2002] SCC 76.

[599] Rimmer, at 91–92

[600] Garland and Smordin, *supra* note 583, at 168–169.

[601] *Id.*, at 169.

[602] *Harvard College*, *supra* note 599 at para. 158.

2.4 Patentability of Biotechnology Inventions in Canada

Considering the specific terms, manufacture and composition of matter used in the definition of invention under Section 2 of the Canadian Patent Act, the Supreme Court concluded that higher life forms do not fit within either of these terms. Regarding the term, manufacture, Bastarache J. stated: 'In my view, while a mouse may be analogised to a "manufacture" when it is produced in an industrial setting, the word in its vernacular sense does not include a high life form.'[603] Further, commenting on the term 'composition of matter' Bastarache J. concluded:

> If the words "composition of matter" are understood this broadly, then the other listed categories of invention, including "machine" and "manufacture", becomes redundant. This implies that "composition of matter" must be limited in some way. Although I do not express an opinion as to where the line should be drawn, I conclude that "composition of matter" does not include a high life form such as the oncomouse.[604]

Emphasising on the need for an unequivocal direction from the Parliament regarding the patentability of higher life forms, the majority observed that the patenting of higher life forms under the current Canadian Patent regime would be a 'radical departure' from the traditional patent regime and would, therefore, require 'an unequivocal direction from Parliament'.[605] Though the majority asserted the fact that the objective of the Patent Act was to promote research as well as encourage and reward scientific and technological developments and innovations, however, it held that Parliament chose an exhaustive definition for invention, and therefore, not everything was intended by Parliament to be patentable.[606]

Referring to Canadian Plant Breeders' Rights Act (PBRA), Bastarache J. noted:

> Far more significant, in my view, is that the passage of the Plant Breeders' Rights Act demonstrates that mechanisms other than the Patent Act may be used to encourage inventors to undertake innovative activity in the field of biotechnology. As discussed above, the Plant Breeders' Act is better tailored than the Patent Act to the particular characteristics of plants, a factor which makes it easier to obtain protection. The quid pro quo is that a narrower monopoly right is granted.[607]

Emphasising the need for legislation similar to Plant Breeder's Rights Act for higher life forms, Bastarache J. observed: 'If a special legislative scheme were needed to protect plant varieties, a subset of higher life forms, a similar scheme may also be necessary to deal with the patenting of higher life forms in general'.[608]

The majority acknowledged the fact that the Patent Act does not explicitly differentiate between lower and higher life forms. However, it explained that making such a distinction 'is nonetheless defensible on the basis of common sense differences between the two'.[609] The majority offered a number of justifications supporting such a distinction:

---

[603] *Id.*, at para. 159.

[604] *Id.*, at para. 160.

[605] *Id.*, at para. 166.

[606] Garland and Smordin, *supra* note 583, at 170.

[607] *Harvard College*, *supra* note 599 at para. 188.

[608] *Id.*, at para. 196.

[609] *Id.*, at para. 199.

First, unlike higher life forms, microorganisms and lower life forms are capable of being produced *en masse* in a similar fashion to the manner in which chemical compounds are prepared. Second, higher life forms have the capacity to display emotion, complexity of behavioural reaction and the predictability of behavioural responses. Third, international instruments such as the World Trade Organization's (WTO) TRIPS and the North American Free Trade Agreement (NAFTA) contain an article whereby members may 'exclude from patentability' certain subject matter, including plants and animals other than microorganisms. This clearly shows that the distinction between higher and lower life forms is widely accepted as valid.[610] Despite making a general distinction between the two life forms, the majority, did not attempt to draw the line between (unpatentable) higher life forms and (patentable) lower life forms.[611]

Referring to a report of the CBAC,[612] the majority noted that 'while the CBAC was in favour of the patenting of higher life forms, that view was also accompanied by a series of recommendations regarding the amendment of the Act, recommendations that clearly required Parliament's input'.[613]

The dissenting opinion in the *Harvard Oncomouse* case has an equal importance in order to understand the Supreme Court's perception regarding patenting of biotechnology in its totality. Expressing the minority opinion on behalf of the dissenting judges, Binnie J. criticized the majority decision for being too narrow in its interpretation of 'composition of matter' and 'manner of manufacture'. He observed:

> "Matter" is a most chameleon-like word. The expression "grey matter" refers in everyday use of "intelligence"-which is about as incorporeal as "spirit" or "mind"… If the oncomouse not composed of matter, what, one might ask, are such things as oncomouse "minds" composed of? The Court's mandate is to approach the issue as a matter (that slippery word in yet another context!) of law, not murine metaphysics.[614]

Commenting on 'common sense' as the basis for distinction between lower and higher life forms, Binnie J. observed: 'With respect, there seems to be as many versions of "common sense" as there are commentators.'[615] He noted that there is no rational dividing line to be drawn between lower life forms and higher life forms in terms of patentability.[616]

From an international perspective, Binne J. referred to the patent practices of different comparable jurisdictions (including the USA, Europe and Japan), which have issued patents on the Harvard Mouse or on related subject matter and contended that Canada was out of step with comparable jurisdictions with similar intellectual property legislations.[617] Further, he maintained that since Canada is signatory of numer-

---

[610] Garland and Smordin, *supra* note 583, at 171.
[611] *Ibid.*
[612] Canadian Biotechnology Advisory Committee (2002).
[613] Garland and Smordin, *supra* note 583, at 171–172
[614] *Harvard college*, *supra* note 599, at 57.
[615] *Id.*, at para. 52.
[616] *Id.*, at paras. 46–56.
[617] *Id.*, at paras. 2, 3 and 33–38.

## 2.4 Patentability of Biotechnology Inventions in Canada

ous international agreements, therefore, it is desirable that comparable jurisdictions with comparable intellectual property legislation arrive at similar legal results.[618]

The minority also quoted from the interim report of the CBAC, *Biotechnology and Intellectual Property: Patenting of Higher Life Forms and Related Issues*, which states: 'the Canadian Patent system is not designed to decide about what uses of technology are permissible nor is the Patent Act designed to prevent dangerous or ethically questionable inventions from being made, used, sold or imported'.[619]

Though the dissenting Judges agreed with the majority that the decision to include higher life forms under patentable subject matter should be left to Parliament, however, they stated that 'neither the Commissioner of Patents nor the court have the authority to declare, in effect, a moratorium on [higher life] form patents until Parliament chooses to act'.[620]

Referring to the *Plant Breeders' Rights Act* 1990 (RSC), Binne J. maintained that the majority erred in drawing a negative inference from the passage in the said Act that plants were not intended by the Parliament to be patentable under the Patent Act 1985 (RSC).[621] He stressed that plant breeder's right and patent rights can live together.[622]

Though the decision of the Supreme Court of Canada in *Harvard College* was welcomed by various social groups including animal rights groups but it had been a big upset for biotechnology industry. In order to alleviate the fear among biotechnology groups that the said decision of the Supreme Court would ruin the biotechnology industry, the CBAC had released an advisory memorandum on 'Higher Life Forms and The Patent Act'.[623] The Committee maintained: 'If the Government of Canada wishes higher life forms to be patentable, it must propose amendments to the Patent Act and gain Parliament's agreement.'[624]

Since the Supreme Court of Canada avoided drawing a strict line between (non-patentable) higher life forms and (patentable) lower life forms, this distinction thus far has been presumably left to the Canadian Patent Office, which currently equates patentable 'lower life forms' with matter that is essentially unicellular.[625] Referring to the *Harvard College* decision, Canadian Manual of Patent Office Practice (MOPOP) maintains:

> The patent office considers the distinction between higher and lower life forms to be, in general, whether the life form is unicellular (lower) or multicellular (higher). The Harvard

---

[618] *Id.*, at paras. 12 and 13.

[619] Canadian Biotechnology Advisory Committee, *Biotechnology and Intellectual Property: Patenting of Higher Life Forms and Related Issues: Interim Report to the Government of Canada Biotechnology Ministerial Coordinating Committee* (CBAC, Otawa, 2001) at 6, cited in Garland and Smordin, *supra* note 583, at 173.

[620] *Harvard College*, *supra* note 599, at para. 114.

[621] *Id.*, at 60.

[622] *Id.*, at 63.

[623] Canadian Biotechnology Advisory Committee (2003).

[624] *Id.*, at 5.

[625] Garland and Smordin, *supra* note 583, at 174.

Decision is interpreted by the Patent Office to mean that animals at any stage of development are non-statutory matter for letters patent, and consequently that fertilised eggs and totipotent stem cells (which have an inherent ability to develop into animals) are included in the higher life form proscription.[626]

*Monsanto Canada Inc. versus Schmeiser*[627] *(Monsanto) Case* The decision of the Supreme Court of Canada in *Monsanto* received great appreciation from biotechnology industry for its relatively broad interpretation of patentable subject matter as compared to *Harvard Oncomouse* case. The majority in *Monsanto* made it clear that even though a higher life form cannot be patented, the patent protection for the genetic material that makes up the life form may extend to protect the higher life form itself.[628] Though the Supreme Court's decision in *Monsanto* signified a departure from the court's line of reasoning in *Harvard College*, however, it was not a reversal of its previous decision in *Harvard College*, at least in the eyes of the court.[629] Since this case has a significant impact on the biotechnology patents in Canada, therefore, a detailed discussion of the case is pertinent.

*Monsanto Canada Inc. versus Schemieser* was a patent infringement case, wherein, a biotechnology company Monsanto Canada Inc. (Monsanto) sued a canola farmer Percy Schmeiser over his unlicensed use of Monsanto's seed which was marketed as Roundup Ready Canola. Roundup Ready Canola contains an inserted gene patented by Monsanto that provides a high level of resistance to the herbicide glyphosate. Glyphosate is an herbicide which ordinarily kills plants. A glyphosate resistance gene, which is the subject of Monsanto's Canadian Patent No. 1,313,830 entitled 'Glyphosate-Resistant Plants', confers resistance to the glyphosate-containing herbicide RoundupTM. Therefore, plants containing the patented gene survive spraying with Roundup. Such a trait allows Roundup Ready Canola users to spray their crops for weeds even after canola plants have emerged.[630] Monsanto introduced Roundup Ready Canola in Canada in 1996 and stipulated certain requirements for the users:

> [A] farmer who wishes to grow Roundup Ready Canola to enter into a licensing arrangement called a Technology Use Agreement ("TUA"). The licensed farmers must attend a Grower Enrolment Meeting at which Monsanto describes the technology and its licensing terms. By signing the TUA, the farmer becomes entitled to purchase Roundup Ready Canola from an authorized seed agent. They must, however, undertake to use the seed for planting a single crop and to sell that crop for consumption to a commercial purchaser authorized by Monsanto. The licensed farmers may not sell or give the seed to any third party, or save seed for replanting or inventory.[631]

---

[626] *Manual of Patent Office Practice* [MPOP] Chapter 17.02.01a "Higher and Lower Life Forms".
[627] *Monsanto, supra* note 541.
[628] Crane (2009).
[629] *Ibid.*
[630] Law and Marles (2004).
[631] *Monsanto, supra* note 541, at para. 11.

## 2.4 Patentability of Biotechnology Inventions in Canada

The license agreement stipulates that new seeds must be purchased every year and an annual licensing fee of C$ 15 for each acre be paid.[632]

Percy Schmeiser had been in his usual practice of planting and saving Canola seeds. In 1997, he sprayed Roundup herbicide to control weeds on a section of his field adjacent to a public road and discovered that some of the canola which had been sprayed had survived. He then tested an additional area of 3 acres to 4 acres of the same field by applying Roundup and found that 60% of the plants survived the spraying. Though the origin of these Roundup resistant plants was not clear, however, it was possible that seed blew onto the Schmeiser's property from neighbouring farms, where Roundup Ready Canola was being cultivated. In 1998, Schmeiser harvested the Roundup resistant canola from the tested area he had sprayed and kept it separate from the rest of his crop. In 1998, Schmeiser used the seeds of the separated roundup resistant canola crops to plant approximately 1000 acres. The samples of canola plants taken from Schmeiser's fields sown in 1998 confirmed after tests that 95–98% of the canola was Roundup resistant.[633] Here, it is worth mentioning that Schmeiser never purchased Roundup Ready Canola nor did he obtain a licence to plant it.[634]

Consequently, in August 1998, Monsanto sued Schmeiser for patent infringement. At the Federal Court Trial Division, the trial judge found that Monsanto's patent was valid and infringed by Schmeiser.[635] The finding of the Trial Court was upheld by the Federal Court of Appeal, which endorsed Monsanto's patent claims as being for 'genes and cells' which are glyphosate-resistant.[636] On appeal, the Supreme Court of Canada by a 5–4 majority held that the patent was valid and that Schmeiser had infringed it.[637]

At the Supreme Court, the majority agreed that Monsanto did not claim protection for genetically modified plants itself but rather for the genes and modified cells that made up the plant.[638] The Monsanto' patent claims were directed to a chimeric gene; an expression vector; a plant transformation vector; various species of plant cells into which the chimeric gene has been inserted; and a method of gene regulating a glyphosate resistant plant.[639]

While acknowledging the Monsanto's patent claims over a gene and a modified cell, Schmeiser contended that extending Monsanto's patent claims to higher life forms such as plants and seeds would be a departure from the *Harvard College*

---

[632] *Id.*, at para. 12.

[633] Law and Marles, *supra* note 631.

[634] *Monsanto*, *supra* note 541, at para. 6.

[635] *Monsanto Canada Inc. v. Schmeiser*, (2001), 202 F.T.R. 78, 12 C.P.R. (4th) 204, [2001] F.C.J. No. 436 (QL), 2001 FCT 256.

[636] *Monsanto Canada Inc. v. Schmeiser*, ([2003] 2 F.C. 165, at para. 40).

[637] *Monsanto*, *supra* note 541.

[638] *Id.*, at para. 17.

[639] *Id.*, at para. 20.

decision in which it was held that plants and seeds were unpatentable life forms.[640] The majority responded this contention by providing different reasons; first, the *Monsanto* case is different from *Harvard College* where the patent refused was for a mammal; second, in *Harvard College* the Patent Commissioner had allowed certain claims regarding plasmid and somatic culture (which were not at issue before the court), somewhat analogous to Monsanto's claims directed to a gene and a cell; and third, all the members of the *Harvard College* case noted in obiter that a fertilised, genetically altered oncomouse egg would be patentable subject matter, regardless of its ultimate anticipated development into a mouse.[641]

The majority concluded that with respect to the patent's validity, the issue whether or not patent protection for the gene and the cell extends to activities involving the plant is not relevant. The majority, however, added that this issue is, however, relevant for examining the factual circumstances in which the infringement has taken place.

Referring to the *Apotex Inc. versus Wellcome Foundation Ltd.*[642] case, the majority opined that Monsanto's patent has already been issued; therefore, the onus is on Schmeiser to show that the Commissioner erred in allowing the patent. In this context, the majority concluded that since Schmeiser has failed to discharge that onus, that patent is valid.[643]

After deciding the validity of Monsanto's patent claims, the court had to decide whether Schmeiser's activities infringed Monsanto's patent, contrary to Sec. 42 of the Patent Act 5. Section 42 provides that 'a patent grants to the patentee the exclusive right, privilege and liberty of making, constructing and using the invention and selling it to others to be used'.[644] After considering the issue in the light of Sec. 42 of the Patent Act, the majority opined that 'the central question on this appeal is whether Schmeiser, by collecting, saving and planting seeds containing Monsanto's patented gene and cell, "used" that gene and cell.'[645]

Emphasising the purpose behind Sec. 42 of the Canadian Patent Act, the majority observed:

> The purpose of Sec. 42 is to define the exclusive rights granted to the patent holder. These rights are the rights to full enjoyment of the monopoly granted by the patent. Therefore, what is prohibited is "any act that interferes with the full enjoyment of the monopoly granted to the patentee."[646]

The majority opined that a defendant has 'used' a patented invention when the defendant has deprived the inventor, either directly or indirectly, of the full enjoyment

---

[640] *Id.*, at para. 21, quoting *Harvard College v. Canada (Commissioner of Patents)*, [2002] 4 S.C.R. 45, 2002 SCC 76 ("Harvard Mouse").

[641] *Id.*, at paras. 22 and 23.

[642] *Apotex Inc. v. Wellcome Foundation Ltd.*, [2002] 4 S.C.R. 153, 2002 SCC 77, at paras. 42–44.

[643] *Monsanto*, *supra* note 541, at para. 24.

[644] R.S.C. 1985, c. P-4, s. 42.

[645] *Monsanto*, *supra* note 541, at para. 28.

[646] *Id.*, at 34, quoting H. G. Fox, The Canadian Law and Practice Relating to Letters Patent for Inventions (4th ed. 1969).

of the monopoly conferred by the patent.[647] In this regard, it was contended that Schmeiser had not 'used' the invention by growing plants because plants were not covered by the claims in Monsanto's patent, and only plant cells containing the modified gene were covered. However, the argument was countered by another argument that the plants were composed of modified plant cells containing the modified genes and therefore, growing the modified plants constitutes use of the invention. The majority relied on existing case law relating to mechanical inventions while supporting the proposition that even if a product as a whole is not covered by a patent, if an important part or component of the product is patented, exploitation of the product may still result in infringement.[648]

Further, the majority also made it clear that possession of an item containing a patented part, at least in commercial circumstances, raises a rebuttable presumption of use, although the absence of intention to employ or gain any advantage from the invention may be relevant to rebutting this presumption.[649] In the light of this reasoning, the majority examined the fact of the case and found that Schmeiser had been in possession of Roundup Ready Canola in a commercial setting, and that he had knowingly selected and planted Roundup Ready Canola in his fields. He used the patented material contrary to the Patent Act by planting and saving seeds, thereby depriving Monsanto of the full enjoyment of its monopoly.[650]

Schmeiser failed to produce credible evidence denying the fact that he used or intended to use the invention to rebut the presumption of use. Though Schmeiser testified that he had not used Roundup to reduce weeds, and therefore did not gain any agricultural advantage from the patented crop, however, the majority found he could have profited from the invention in the future by using Roundup or by selling the seed. Nevertheless, since Schmeiser had not actually gained any profits from his use of the invention, the majority held that Monsanto was not entitled to an accounting of profits, despite the finding of infringement.[651]

The minority concluded that the patent claims in Monsanto cannot be interpreted to extend patent protection over whole plants and that there was no use leading to infringement.[652] The minority explained the scope of patent claims:

> It is clear from the specification that Monsanto's patent claims do not extend to plants, seeds, and crops. It is also clear that the gene claim does not extend patent protection to the plant. The plant cell claim ends at the point where the isolated plant cell containing the chimeric gene is placed into the growth medium for regeneration. Once the cell begins to multiply and differentiate into plant tissues, resulting in the growth of a plant, a claim should be made for the whole plant. However, the whole plant cannot be patented. Similarly, the method claim ends at the point of the regeneration of the transgenic founder plant

---

[647] Law and Marles, *supra* note 631, at 45.
[648] *Ibid.*
[649] *Ibid.*
[650] *Ibid.*
[651] *Ibid.*
[652] *Monsanto*, supra note 541, at para. 111.

but does not extend to methods for propagating that plant. It certainly does not extend to the offspring of the regenerated plant.[653]

The minority agreed that Monsanto's claims as to products and process were valid, and these valid claims were restricted to genetically modified chimera genes and cells in the laboratory prior to regeneration and the attended process for making the genetically modified plant.[654]

Regarding the infringement issue, the minority took the position that 'use' of an invention under Sec. 42 of the Patent Act should be constrained by the subject-matter of the claims. Rejecting the majority's mechanical analogy, the minority explained that existing case law on 'use' and analogies to mechanical inventions are not helpful in this context because of the unique ability of biological organisms to self-replicate.[655] The minority concluded:

> In the result, the lower courts erred not only in construing the claims to extend to plants and seed, but in construing "use" to include the use of subject matter disclaimed by the patentee, namely the plant. The appellants as users were entitled to rely on the reasonable expectation that plants, as unpatentable subject matter, fall outside the scope of patent protection. Accordingly, the cultivation of plants containing the patented gene and cell does not constitute an infringement. The plants containing the patented gene can have no stand-by value or utility as my colleagues allege. To conclude otherwise would, in effect, confer patent protection on the plant.[656]

The majority in Monsanto maintains that many biotechnology inventions will only receive the full benefit of patent protection if the scope of the patent extends to genetically modified organism as a whole. However, the majority asserted the fact that it had only interpreted the Patent Act as it perceives, and Parliament remains free to pursue any amendments.[657] As the things stands of now, Canada still maintains a distinction between 'higher' and 'lower' life forms in the absence of any clear direction from the Parliament. *Canadian Manual of Patent Office Practice* (MOPOP) contains a list of (patentable) lower life forms and (non-patentable) higher life forms:

> Lower life forms include: microscopic algae; unicellular fungi (including moulds and yeasts); bacteria; protozoa; viruses; transformed cell lines; hybridomas; and embryonic, pluripotent and multipotent stem cells.
> Higher life forms include: animals, plants, seeds, mushrooms, fertilized eggs and totipotent stem cells.[658]

---

[653] *Id.*, at para. 130.
[654] *Id.*, at paras. 138 and 139.
[655] Law and Marles, *supra* note 631.
[656] *Monsanto*, *supra* note 541, at para. 160.
[657] Law and Marles, *supra* note 631, at 47.
[658] Manual of Patent Office Practice (MPOP), *supra* note 627.

### 2.4.1.2 Other Statutory Requirements for Patenting

In addition to the subject matter requirement, only those inventions that are new, useful and non-obvious can be patented.

**Novelty** Sec. 2 of the Canadian Patent Act prescribes that to be patentable an invention must be new i.e. an invention must not have been revealed previously to the public (subject to a 1-year grace period in Canada).[659] The Canadian Patent Act contains the concept of absolute novelty i.e. the invention must not have been described or claimed in a previously filed third party Canadian patent application and must not have been described previously publicly by a third party anywhere in the world. An exception to the absolute novelty provision is disclosure to the public by an applicant or by a person who obtained knowledge of the invention from the applicant, directly or indirectly. Such a disclosure can be made up to a year before the Canadian filing date.[660]

**Utility** The utility requirement also stems from Sec. 2 of the Canadian Patent Act which prescribes that an invention to be patentable must be useful.[661] A useful invention is one that can be put into practice. The requirement of utility under Canadian Patent Act is relatively low as compared to that in the USA which requires specific, substantial and credible utility. In Canada, an invention is required to do what it is supposed to do in order to meet the utility criteria; it does not need to work all of the time nor be better than existing product and processes, but it must work.[662] The Supreme Court of Canada in *Apotex Inc. versus Wellcome Foundation Ltd.*[663] stated that an invention will be considered useful even if it is uncertain that it will do what it is supposed to do, if there exist good reasons to believe that the invention will work as claimed. For instance one can claim that a medicine is useful based on animal studies even though the medicine has yet to be proven effective in humans.[664] Though the utility requirement under Canadian Patent Act is easy to meet, nevertheless, it does limit the scope of protection by excluding methods that would not be useful. Meaning thereby, utility requirement is used to check the patentable subject matter requirement under Canadian Patent Act. Some functional support must be provided for claims in a biotechnology patent.[665]

**Non-obviousness** S. 28.3 of the Canadian Patent Act describes the requirement of non-obviousness: 'The subject-matter defined by a claim in an application for a pat-

---

[659] Patent Act, R.S.C. 1985, c. P-4, ss. 2, 28.2(1)(a).
[660] *Id.*, s. 28.2 (1)(a)-(d).
[661] *Id.*, s. 2.
[662] Gold (2009b).
[663] *Apotex Inc. v. Wellcome Foundation Ltd.*, [2002]4 S.C.R. 153(S.C.C.).
[664] Gold and Knoppers, *supra* note 663, at 21.
[665] Nador, *supra* note 543.

ent in Canada must be subject-matter that would not have been obvious on the claim date to a person skilled in the art or science to which it pertains,...'[666] The classic test is whether a person with all the relevant knowledge in the world but without any creativity would have come up with the invention.[667]

**Written Description, Enablement and Best Mode** S. 27.3 of the Canadian Patent Act contains provisions regarding written description, enablement and best mode. It states:

> The specification of an invention must
> 
> a. correctly and fully describe the invention and its operation or use as contemplated by the inventor;
> b. set out clearly the various steps in a process, or the method of constructing, making, compounding or using a machine, manufacture or composition of matter, in such full, clear, concise and exact terms as to enable any person skilled in the art or science to which it pertains, or with which it is most closely connected, to make, construct, compound or use it;
> c. in the case of a machine, explain the principle of the machine and the best mode in which the inventor has contemplated the application of that principle; and
> d. in the case of a process, explain the necessary sequence, if any, of the various steps, so as to distinguish the invention from other inventions.[668]

Further, applicants can also now rely on a reference in a specification to a deposit of biological material to supplement the written description of an invention. Where an invention is biological material or where the invention relies on biological material, words alone may not be sufficient to fully describe the invention and, therefore, to satisfy the disclosure requirements of the Act. In such cases, access to the biological material may also be necessary.[669]

With regard to the biological materials, Canadian Patent Rules enables that patent specifications which disclose nucleotide and/or amino acid sequences which are not prior art sequences, must now include a 'sequence listing' and be accompanied by a copy of the listing in electronic format i.e. on diskette. A 'sequence listing' includes the actual sequence(s) and associated information and is part of the description section of the specification.[670]

## 2.4.2 Comparison of Canada with the USA and Europe

Canada has adopted a unique position over patenting of higher life forms. Though its policies are largely influenced by the USA, it is moving cautiously with regard to biotechnology patents. Although it gives importance to the economic signifi-

---

[666] Patent Act, R.S.C. 1985, c. P-4, s. 28.3.
[667] *Supra* note 664.
[668] Patent Act, R.S.C. 1985, c. P-4, s. 27.3.
[669] *Id.*, s. 28.1.
[670] *Patent Rules*, s. 111.

cance of biotechnology patents, however, it has been a little slow in granting patents on life forms as compared to the USA. The general public perception favours the biotechnology patents and morality and public order has not restrained the biotechnology developments. However, Canadian approach towards patenting of higher life forms had been influenced by the persistent opposition of Harvard Oncomouse patent in Europe (though the opposition was set to rest in Europe). Canadian courts have not been interpreting the provisions of the Canadian Patent Act so widely to accommodate higher life forms despite having an essentially similar patent law to the USA. Further, Canada does not have specific legislation regarding biotechnology such as Biotechnology Directive in Europe. The Canadian Patent Act does not contain an elaborative list of patentable and non-patentable subject matter such as EPC. Like the USA, Canadian courts are giving shape to the biotechnology patents by interpreting the relevant provisions of Canadian Patent Act to provide various exclusions from patentability and accommodate new subject matters under Canadian Patent Act. Canada lacks specific guidelines regarding the patentability criteria such as USPTO guidelines on utility and guidelines of Trilateral Project (USPTO, EPO and JPO). Gene patents are being granted by the Canadian Patent Office; however, these patents are not contested in the courts on a regular basis to give a clear picture on various aspects of gene patenting. It is yet to be seen how courts in Canada would interpret gene patent cases.

## 2.5 Patentability of Biotechnology Inventions in India

There has been relatively slow growth of biotech patents in India as compared to the USA and Europe. Numerous factors have been responsible for the slow growth. Among them some important factors are: first, product patents on substances capable of use as medicine, drug or food could not be obtained from the Indian Patent Office prior to 1 January 2005, when India complied with the implementation of TRIPS-required amendments to its Patents Act; second, even the process patent had always been available regarding such substances but the policy of Indian Patent Office had been not so encouraging regarding the grant of patents on processes that produced a live product till *Dimminaco A.G. versus Controller of Patents, Designs & Trade Marks*[671] *(Dimminaco)* decision and third, there had not been explicit provisions regarding biochemical, biotechnological and microbiological processes till 2002 amendment to Indian Patent Act.[672]

Three major amendments (1999, 2002 and 2005) have taken place in order to bring the Indian Patent Act 1970 in full conformity with the obligations mandated under TRIPS. These amendments have paved the way for biotechnology patents.

---

[671] *Dimminaco A.G. v. Controller of Patents, Designs & Trade Marks*, (2002) I.P.L.R. July 255, 269 (Calcutta H.C.).
[672] Mueller (2008).

These amendments were accomplished during a 10-year transition period (1995 through 2004) by three acts amending the 'principal' act i.e. the Patents Act, 1970.[673]

The Patents (Amendment) Act, 1999,[674] introduced exclusive marketing rights (EMR) and established mail box applications for patents for pharmaceuticals and agrochemicals from 1 January 1995.[675] The Patents (Amendment) Act, 2002,[676] implemented the TRIPS-required 20-year patent term, reversal of the burden of proof for process patent infringement, and modifications to compulsory licensing requirements. The 2002 Act, for the first time extended the scope of patentable subject matter to accommodate biotechnological inventions by referring biotechnological and microbiological processes. The Act clarified that 'chemical processes', which were already considered patentable but given a patent term of only 5–7 years, included 'biochemical, biotechnological, and microbiological processes'.[677] Finally, through Patents (Amendment) Act, 2005,[678] India has given full effect to pharmaceutical product patents.[679]

Before discussing the relevant statutory provisions regarding biotechnology inventions under the current Indian Patent Act 1970 (as amended in 1999, 2002 and 2005), it becomes pertinent to discuss *Dimminaco* case which opened the gate for patents on life forms. This decision has a significant impact on the amendments followed.

### 2.5.1 Dimminaco Case: Paving the Way for Biotechnology Patents in India

Indian Commentators see this case as important as *Diamond versus Chakrabarty*[680] in the USA.[681] In *Dimminaco*, on 15 January 2002, the Calcutta High Court held that a process for preparation of a vaccine, the end product of which contained a live virus, was an 'invention' eligible for protection under the Patents Act.[682] The High Court has overturned the long-standing policy of the Indian Patent Office to refuse such process claims and paved the way for biotechnology patents in India.[683]

Dimminaco's patent application was directed to a process for preparing a vaccine to protect chickens against infectious bursitis. Patent examiner refused the grant of

---

[673] *Ibid.*
[674] The Patents (Amendment) Act, 1999 (17 of 1999).
[675] Jauhar and Narnaulia (2010).
[676] The Patents (Amendment) Act, 2002 (38 of 2002).
[677] Mueller, *supra* note 673.
[678] The Patents (Amendment) Act, 2005 (15 of 2005).
[679] Mueller, *supra* note 673.
[680] 447 U.S. 303 (1980).
[681] Mueller, *supra* note 673.
[682] *Diminaco*, (2002) I.P.L.R. July 255 (Culcutta H.C.), at 259.
[683] Mueller, *supra* note 673.

## 2.5 Patentability of Biotechnology Inventions in India

patent and the refusal was upheld by the Assistant Controller.[684] The Patent Office took the position that although the statute defined a potentially patentable invention as including a new and useful 'art, process, method or manner of manufacture', a process that produced a living end-product (as here, the vaccine containing a live virus) was not within the statutory classifications.[685] The Assistant Controller looked into the previous record of patent offices and found that there was no record showing patents granted for a process of preparing a living organism. The Controller referred to 'Iyengar' Committee which recommended that 'invention' should be defined narrowly.[686] Dimminaco, on the other hand, contended that since the Patent Office had neither cited any anticipatory prior art against the claimed process, nor had it questioned the utility of the end-product vaccine, therefore, the process should be held a patentable 'manner of manufacture' coming within the Patent Act's enumerated categories of inventions.[687]

When the case reached to the Calcutta High Court, the bone of contention was whether the appellant's claim comes under the scope of 'manner of manufacture' under 2(i)(j) of the Patents Act, 1970. Considering the scope of 'manner of manufacture', the court emphasised, the dictionary meaning of 'manufacture' or its usage 'in the particular trade or business [at issue] must be accepted'.[688] The court further opined that in order to decide whether a particular process of manufacture 'ought to be patented or not, "one of the most common test[s] is the vendibility test."'[689] This test is satisfied 'if the invention results in the production of some vendible item [,]... something which can be passed on from one man to another upon the transactions of purchase and sale.'[690]

In the light of the dictionary meaning of 'manufacture', the court concluded that in the present case the claimed method was patentable invention because '[t]he dictionary meaning of the word manufacture does not exclude the process of preparing a vendible commodity which contains a living substance.'[691] The court observed that since the Indian Patent Act 1970 did not provide the definition of the term 'manufacture', therefore, the dictionary meaning must be accepted.[692]

While overturning the decision of Assistant Controller's, the court referred to Indian Patent Nos 183034 and 183035, both directed to the process of preparing therapeutic preparations for treatment of diarrhoea, where the end products comprised live *Lactobacillus* (L.) *reuteri* cells.[693] The court concluded 'that patent has

---

[684] *Diminaco*, (2002) I.P.L.R. July 255 (Calcutta), at 258, 270.
[685] *Id.*, at 258, 259, 261.
[686] Mueller, *supra* note 673.
[687] Diminico, (2002) I.P.L.R. July 255 (Calcutta), at 261, 262.
[688] *Id.*, at 268.
[689] *Ibid.*
[690] *Ibid.*
[691] *Id.*, at 269.
[692] *Ibid.*
[693] *Id.*, at 266–268.

been granted by the authorities in cases where end-product contains living virus.'[694] Though the court did not engage in a detailed discussion regarding patent policy, however, it quoted the Indian Supreme Court's 1982 statement in *M/s. Bishwanath Prasad Radhey Shyam versus M/s. Hindustan Metal Indus*[695] that 'the object of the patent law is to 'encourage scientific research, new technology and industrial progress'.[696] Finally, the court in *Dimminaco* concluded that the Assistant Controller erred in rejecting Dimminaco's application on the ground that it could not be called a 'manner of manufacture' because it involved a living virus in the end-product.[697]

Here, it is noteworthy that the *Dimminaco* case was concerned with the process patent over the process of manufacture, it, therefore, only allowed for patenting of a process resulting in a product with living matter, not over the living matter itself. Further, the case was also fundamentally different from that of a gene patenting, which involves several complex issues such as impediment to research, access to genetic testing, gene as information etc.[698]

Though, the case was decided in favour of the appellant, the court reserved its right to decide such cases on factual basis after detailed scrutiny of the facts of each case.[699]

## 2.5.2 Statutory Provisions Regarding Biotechnological Inventions Under the Current Patent Act 1970 (as Amended in 1999, 2002 and 2005)

### 2.5.2.1 Biotechnological Inventions as Patentable Subject Matter

The Patent Amendment Act 2002 came into effect subsequent to *Dimminaco* decision. The amendment has changed the definition of 'invention'.

Before the Patent Amendment Act 2002, Sec. 2(j) contained a complex definition of invention as being art, product or process, method or manner of manufacture; machine, apparatus or other article; substance produced by manufacture, including any useful improvements on the said.[700] This definition has been substituted by a new simplified definition of invention by the Patent Amendment Act 2002. The new definition under Sec. 2(1)(j) states: '"invention" means a new product or process involving an inventive step and capable of industrial application.'[701]

---

[694] *Id.*, at 270.

[695] *M/s. Bishwanath Prasad Radhey Shyam v. M/s. Hindustan Metal Indus.*, A.I.R. 1982S.C. 1444.

[696] *Dimminaco*, I.P.L.R. July 255 (Calcutta), at 269.

[697] *Id.*, at 269–70.

[698] Jauher and Narnaulia, *supra* note 676, at 60.

[699] *Ibid.*

[700] *Id.*, at 61.

[701] Sec. 2(1)(j) of The Patent Act, 1970 (Substituted by the Patent (Amendment) Act 38 of 2002, Sec. 3, for clause (j) (w.e.f. 20-5-2003).

Furthermore, Patent Amendment Act 2005 repealed Sec. 5 of the parent Patent Act 1970.[702] Sec. 5 provided for only process patents on chemical processes as well as substances intended for or capable of being used as food or medicine. In this context, if the Hon'ble High Court's intention in *Dimminaco* is properly read, it becomes apparent that the doors have been opened for patenting products with living organisms with both legislative as well as judicial sanction.[703]

*Back to the statutory framework,* for a patent to be granted in India it should not be covered in the negative list in Section 3 which provides an extensive list of what are not inventions (exclusions) under the Indian Patent Act.[704]

One of the exclusions, relevant to biotechnological invention is found in Sec. 3(c) of the Indian Patent Act which prohibits patents on discovery of any living thing occurring in nature.[705] Further, Sec. 3(j) provides that plants and animals in whole or any part thereof including seeds, varieties, species, and essentially biological processes for production and propagation of plants and animals are not patentable subject matter.[706] By implication of Sec. 3 genetically modified multicellular organisms including plants, animals and human beings, and their parts are not patentable in India.[707]

Sec. 3(j) of the Patent Act, 1970, however, allows patents for microorganisms and microbiological processes.[708] Therefore, microorganisms, other than the ones discovered from the nature, may be patentable.

The *Manual of Patent Practice and Procedure, 2005,* provides that biological materials such as recombinant DNA, plasmids and processes of manufacturing thereof are patentable if they are produced by substantive human intervention.[709] Gene sequences and DNA sequences with disclosed function are considered patentable in India.[710] The processes for cloning human beings or animals, processes for modifying the germ line, genetic identify of human beings and animals, uses of human or animal embryos for any purpose are not patentable as they are against public order and morality.[711] Sec. 3(i) contains a list of non-patentable inventions in India which includes 'any process for the medical, surgical, curative, prophylactic, diagnostic, therapeutic or other treatment of human beings or any process of a similar treatment of animals to render them free of disease or to increase their economic

---

[702] Rep. by the Patents (Amendment) Act 15 of 2005, Sec. 4 (w.r.e.f. 1-1-2005).

[703] Jauher and Narnaulia, *supra* note 676, at 61.

[704] Sec. 3 of the Patents Act, 1970 (as amended in 1999, 2002 and 2005).

[705] Sec. 3(c) of the Patents Act, 1970 [Ins. by Patent (Amendment) Act 38 of 2002, Sec. 4 (w.e.f. 20-5-2003)].

[706] Sec. 3(j) of the Patents Act, 1970 [Ins. by Patent (Amendment) Act 38 of 2002, Sec. 4 (w.e.f. 20-5-2003)].

[707] Kankanala, *supra* note 272.

[708] *Supra* note 693.

[709] Annexure 1, Examination Guidelines for Patent Applications relating to inventions in the Field of Chemicals, Pharmaceuticals and Biotechnology, *Manual Patent Practice and Procedure,* 2005.

[710] *Ibid.*

[711] *Id.*, citing Sec. 3(c), Indian Patent Act, 1970, as amended in 2002.

value or that of their products.'[712] As genetic therapies are therapeutic and are meant for rendering an animal or human being free of disease, they are not patentable in India.[713] Plant varieties are provided protection in India under the provisions of the Protection of Plant Varieties and Farmers' Rights Act, 2002.

### 2.5.2.2 Other Statutory Requirements Under Indian Patent Act for Patenting

**Novelty** Section 2(1)(j) of the Indian Patents Act requires that the invention must be new,[714] that is, it must be different from 'prior art'. Sec. 2(l) describes 'new invention' as:

> ...[a]ny invention or technology which has not been anticipated by publication in any document or used in the country or elsewhere in the world before the date of filing of patent application with complete specification., i.e. the subject matter has not fallen in public domain or that it does not form part of the state of the art.[715]

The *Manual of Patent Office Practice and Procedure (MPPP), 2011*, explains that 'a prior art will be considered as anticipatory if all the features of the invention are present in the cited art.'[716] It further states that 'the prior art should disclose the invention either in explicit or implicit manner.[717]

**Inventive Step (Non-obviousness)** According to Sec. 2(1)(ja) of the Indian Patent Act, 'inventive step' means a feature of an invention that involves technical advance as compared to the existing knowledge or having economic significance or 'both' and that makes the invention not obvious to a person skilled in the art.'[718]

It is apparent from Sec. 2(1)(ja) that in order to fulfil the requirement of inventive step, an invention should either have a technical advance or economic significance. Moreover, the invention should not be obvious to a person with ordinary skill in the art. The actual meaning and scope of technical advance and economic significance is difficult to be ascertained due to dearth of case law and legislative history. However, it can be inferred from the guidance provided by the *Manual of Patent Practice and Procedure* that isolated gene sequences will be considered to have an inventive step in the light of their naturally existing counterparts.[719]

---

[712] Sec. 3 of the Patents Act 1970, as amended by the Patents (Amendment) Act 38 of 2002.

[713] Kankanala, *supra* note 272, at 27.

[714] *Supra* note 702.

[715] Sec. 2(l) Sec. 2(1)(l) of the Patents Act, 1970, substituted by the Patent (Amendment) Act 15 of 2005, Sec. 2(g), for clauses (l) and (m) (w.r.e.f. 1-1-2005).

[716] *MPPP, 2011*, Chapter 08.03.02.

[717] *Ibid.*

[718] Sec. 2(1)(j)(a) of the Patents Act, 1970, substituted by the Patent Amendment Act 38 of 2002s. 3, for clause (j) (w.e.f. 20-5-2003) and Patent (Amendment) Act 15 0f 2005, Sec. 2 (f), for clause (ja) (w.r.e.f. 1-1-2005).

[719] Kankanala, *supra* note 272, at 96.

## 2.5 Patentability of Biotechnology Inventions in India

Since Indian law recognises economic significance as a determining factor for inventive step, therefore, it would be easy for gene-based inventions to satisfy this requirement. This is because most gene sequences and related inventions have great significance in the biopharma sector because of the multitude of uses to which they can be put. It would therefore be easy for gene-based inventions to pass the test of inventive step in India than in USA or Europe, which do not give primary importance to commercial value of an invention.[720]

**Industrial Applicability** Sec. 2(1)(j) of the Patents Act, 1970, mandates that an invention is patentable in India, if it is capable of being industrially applicable.[721] Sec. 2(ac) of the said Act provides the meaning of 'capable of industrial application' as, '"capable of industrial application", in relation to an invention, means that the invention is capable of being made or used in an industry.'[722] The provision is similar to the EPC. The *Manual of Patent Practice and Procedure, 2005,* provides that Industrial application of an invention is determined by confirming if the invention:

Can be made
Can be used in at least one field of activity, and
Can be reproduced with the same characteristics as many times as necessary.[723]

Since gene sequences and related inventions can be made and used in an industry and can be reproduced as many times as required, therefore, they would satisfy the industrial applicability requirement in India. Though the 'manufacture' and 'reproducibility' provisions are quite straight forward, it would be interesting to see how the Indian Patent Office and the courts interpret the 'use' criteria.[724]

As per the guidelines for examining biotechnology inventions in the *Manual of Patent Practice and Procedure, 2005*, gene sequences and DNA sequences whose functions are not disclosed do not satisfy the requirement of industrial applicability.[725] Similar to the USA and Europe, function of a gene should be known in order to receive patent protection in India.[726] Moreover, the Indian Patent Act provides that new uses of a known invention are not patentable.[727] So in accordance with the Indian Patent Act, it can be inferred that discovery of a new function for a known gene would not get patent protection in India.[728]

*MPPP, 2011*, throws light upon the assessment of the industrial applicability of an invention. It provides:

---

[720] *Ibid.*

[721] *Supra* note 702..

[722] Sec. 2(1)(ac) of the Patents Act 1970, substituted by the Patents (Amendment) Act 38 of 2002, Sec. 3, for clause (a) (w.e.f. 20-5-2003).

[723] *MPPP*, 2005, Chapter II, Page 13, Para. 2.4.

[724] Kankanala, *supra* note 272, at 40.

[725] 7.0 Biotechnology Inventions, Annexure I, *Manual of Patent Practice and Procedure, 2005.*

[726] Kankanala, *supra* note 272, at 41.

[727] Sec. 3(d) of the Patents Act 1970 as amended in 2005.

[728] *Supra* note 725.

....Typically, the specification explains the industrial applicability of the disclosed invention in a self-evident manner. Usually industrial applicability is self-evident. If it is not, a mere suggestion that the matter would be industrially applicable is not sufficient. A specific utility should be indicated in the specification supported by the disclosure. For example, indicating that a compound may be useful in treating unspecified disorders, or that the compound has—useful biological properties, would not be sufficient to define a specific utility for the compound. The specific usefulness has to be indicated.[729]

**Written Description, Enablement and Best Mode** Section 10 of the Indian Patent Act contains detailed provisions relating to specification. Section 10(4)(a) of the Indian Patent Act provides India's version of an enablement requirement. It requires a complete (as opposed to provisional) specification shall 'fully and particularly describe the invention and its operation or use and the method by which it is to be performed.'[730] Further, similar to the USA, Indian Patent Act 1970 contains a best mode provision. Section 10(4)(b) of the said Act provides that the complete specification of an Indian patent shall 'disclose the best method of performing the invention which is known to the applicant and for which he is entitled to claim protection.'[731]

One of the notable features of the Indian patent application regarding biotech patents is that any claim in the application directed to genetic material must be supported by a disclosure of function. In this regard, *the MPPP, 2005*, provides that '[g]ene sequences, [and] DNA sequences without having disclosed their functions are not patentable for lack of inventive step and industrial application.'[732]

The Indian Patent Act 1970 contains specific provisions regarding the patent applications involving biological materials. Section 10(4)(d)(ii) stipulates that if an Indian patent application discloses a 'biological material' that 'may not be described in such a way as to satisfy' the enablement and best mode requirements described above, 'and if such material is not available to the public,'[733] the biological material must be deposited in an international depository authority (IDA) under the Budapest Treaty.[734] Certain other conditions must also be satisfied for biological deposits.[735] Among them the relevant condition regarding applications disclosing biological materials is prescribed under Section 10(4)(d)(ii)(D), which provides that

---

[729] *MPPP, 2011*, Chapter 08.03.04.

[730] Sec. 10(4)(a) of the Patents Act, 1970.

[731] Sec. 10(4)(b) of the Patents Act, 1970.

[732] *MPPP, 2005*, 7.0, at 142 (examination guidelines pertaining to biotechnological inventions).

[733] 1970 Patents Act, No. 39, Sec. 10(4)(d)(ii). The Act was amended in 2005 to reflect India's accession to the Budapest Treaty; the prior version of the Act required deposit of biological material 'to an authorized depository institution as may be notified by the Central Government in the Official Gazette.'

[734] 2005 Patents (Amendment) Act, No. 15, Sec. 8(b)(i) (amending 1970 Patents Act, § 10(4)(d)(ii)).

[735] 1970 Patents Act, Sec. 10(4)(d)(ii)(A)-(D).

## 2.5 Patentability of Biotechnology Inventions in India

the applicant must 'disclose the source and geographical origin of the biological material in the specification, when used in an invention.'[736]

India has only one IDA which serves all of India. This IDA is the Microbial Type Culture Collection (MTCC) at the Institute of Microbial Technology (IMTECH) in Chandigarh, India. However, this IDA is of limited scope as it is not equipped to accept deposits of 'cell lines, cyanobacteria, viruses etc.', the deposit of such biological materials would be made in IDAs which are situated outside of India. This would be an administratively cumbersome and costly affair for domestic biotechnology firms.[737]

**Morality** Similar to Europe, India also recognises public order and morality exclusion from patentability. Section 3(b) of the Indian Patents Act provides that 'an invention, the primary or intended use or commercial exploitation of which would be contrary to public order and morality or which causes serious prejudice to human, animal or plant life or health or to the environment' is not patentable.[738] The *Manual of Patent Practice and Procedure, 2005,* provides that any biological material and method of making the same which is capable of causing serious prejudice to human, animal or plant life or health or to the environment including the use of those that would be contrary to public order and morality are not patentable, such as terminator gene technology.[739] The Manual also explains that the processes for cloning human beings or animals, processes for modifying the germ line, genetic identity of human beings or animals, uses of human and animal embryos for any purpose are not patentable as they are against public order and morality.[740]

It is apparent from the Indian Patent Act, 1970, that India strongly prohibits patents for genes and gene-based inventions based on morality and public order.[741] Further, deep-rooted moral, cultural and religious beliefs in India set a litmus test for gene related inventions in order to pass the patentability criteria. So morality and public order plays an important role in determining patentability of gene-based inventions in India.[742] Indian Patent manual 2011 echoes the same concern by mentioning that, 'An invention, the primary or intended use of which is likely to violate the well accepted and settled social, cultural, legal norms of morality, e.g. a method for cloning of humans.'[743]

---

[736] Sec. 10(4)(d)(ii)(D) of the Patents Act 1970, substituted by the Patents (Amendment) Act 15 of 2005, Sec. 5(b)(ii), for sub-clause (A) (w.r.e.f. 1-1-2005).

[737] Mueller, *supra* note 673.

[738] Sec. 3(b) of the Patents Act, 1970, substituted by the Patent Amendment Act 38 of 2002, Sec. 4, for clause (b) (w.e.f. 20-5-2003).

[739] Para. 7.0 Annexure I, Examination guidelines for patent applications relating to inventions in the field of chemicals, pharmaceuticals and biotechnology, *Manual Patent Practice and Procedures*.

[740] *Ibid.*

[741] Kankanala, *supra* note 272, at 138.

[742] *Ibid.*

[743] *MPPP, 2011*, Chapter 8 Para. 03.05.02 page 80–81.

## 2.5.3 Status of Biotechnology Patent in India

Subsequent to the amendments made to the parent Indian Patent Act in 1999, 2002 and 2005, there has been a phenomenal growth in the number of patent applications and consequently an increase in the number of patents granted in the field of biotechnology.[744] An assessment of the annual reports of the Indian Patent Office (2007 Annual Report)[745] reveals that from 2000–2001, where four applications were submitted and no patents were granted, the number of applications has been steadily increasing.[746] In 2003–2004, 23 patent applications had been filed, though no patents were granted. During 2004–2005, there was a phenomenal increase in the number of applications; 73 biotechnology patents were granted in that year alone. The tally for 2007–2008 as in available report stands at 1950 applications and 314 grants.[747]

Despite the increase in the biotechnology patents, India seems to be reluctant to grant patent on life forms. This can be inferred from India's insistence upon substantial review of Article 27.3 of TRIPS and support for the African group's proposal[748] on review of the said article presented in 1999 (which suggests that 'patents on life should be prohibited, including those on microbiological processes)'.[749] India has also adopted a more restrictive European approach on patents by opting to utilize the morality clause in TRIPS.[750]

## References

Boettiger Sara and Bennet Alan B (2006) Bayh-Dole: If We Know Then What We Know Now. 24 *Nature Biotechnology* 24: 320
Karen I. Boyd (1997) Non-obviousness and the Biotechnology Industry: A Proposal for a Doctrine of Economic Obviousness Berkeley Technology Law Journal 12: 311. Available via West Law. www.international.westlaw.com. Accessed 15 May 2011
Bryan Erin (2009) Gene Protection: How Much is too much? Comparing the Scope of Patent Protection for Gene Sequences Between the United States and Germany. Journal of High Technology Law 9: 52
Bud Robert (1991) Biotechnology in Twentieth Century. Social Studies of Science 21: 422
Dan L. Burk (1991) Biotechnology and Patent Law: Fitting Innovation to Procrustean Bed. Rutgers Computer and Technology Law Journal 17: 1. Available via West Law. www.international.westlaw.com. Accessed 15 May 2011
Cain Brian (2003a) Legal Aspects of Genetic Technology. Sweet & Maxwell Ltd, London p. 122
Cain Brian (2003b) Legal Aspects of Genetic Technology. Sweet & Maxwell Ltd, London

---

[744] Jauher and Narnaulia, *supra* note 676, at 61.

[745] Indian Patent Office, Annual Report 2007–08, available at http://ipindia.gov.in/cgpdtm/Annual Report_English_2007-2008.pdf (last visited on September 20, 2011).

[746] *Ibid.*

[747] *Ibid.*

[748] WT/GC/W/302.

[749] Jauher and Narnaulia, *supra* note 676, at 61.

[750] *Ibid.*

Canadian Biotechnology Advisory Committee (2002) Patenting of Higher Life Forms and Related Issues. CBAC, Ottawa

Canadian Biotechnology Advisory Committee (2003) Advisory Memorandum: Higher Life Forms and the Patent Act, Ottawa: Canadian Advisory Committee

Conley John M. (2009) Gene Patents and the Product of Nature Doctrine. Chicago -Kent Law Review 84: 115

Conley John & Vorhaus Dan (2011a) Pig Return to Earth: Federal Circuit Reinstates Most-But Not All-of Myriad's Patents. Genomics Law Report web blog post, 31 July 2011 http://www.genomicslawreport.com/index.php/2011/07/31/pigs-return-to-earth-federal-circuit-reinstates-most-but-not-all-of-myriads-patents/. Accessed 30 August 2011

Conley John & Vorhaus Dan (2011b) Prometheus Returns to the Supreme Court, Medical Patent Speculation Intensifies: Genomics Law Report web blog post, (12 June 2011) http://www.genomicslawreport.com/index.php/2011/06/22/prometheus-returns-to-the-supreme-court-medical-method-patent-speculation-intensifies/. Accessed 31 August 2011

Crane Adam (2009) Of Mice and Man: Patentability of Genetic material and the Protection of Intellectual Property Rights. Dalhousie Journal of Legal Studies 18: 102

Davis Paula K., Kelley James J., Caltrider Steven P. et al. (2005) ESTs stumble at the utility threshold. Nature Biotechnology 23: 1227

Demaine Linda J. & Fellmeth Aaron Xavier (2002) Reinventing the Double Helix: A Novel and Non-Obvious Reconceptualization of the Biotechnology Patent. Stanford Law Review 55: 318 (2002)

Dutfield Graham (2009) Intellectual Property Rights and the Life Science Industries- Past, Present and Future 2nd edn. World Scientific Co. Pte. Ltd, Singapore

Eisenberg Rebecca S. (2002) Why the Gene Patenting Controversy Persists. Academic Medicine 77: 1382

Eisenberg Rebecca (2003) Patenting genome research tools and the law. C. R. Biologies 326: 1116

Eisenberg Rebecca (2006) The story of Diamond v. Chakrabarty: technological change and the subject matter boundaries of the patent system. In: Ginsberg Jane and Dreyfuss Rochelle Cooper (eds.), Intellectual Property Stories. Foundation Press, New York p. 349

Garland Steven B. and Smordin Sandee (2003) The Harvard Mouse Decision And Its Future Implications. Canadian Business Law Journal 39: 162

Gold E. Richard (2009a) The Ethics of Biotechnological Intellectual Property. In: Gold E. Richard & Knoppers Bartha Maria (eds.), Biotechnology IP & Ethics. Lexis Nexis Canada Inc., Markham, Ontario 22

Gold Richard (2009b) The Ethics of Biotechnological Intellectual Property. In: Gold E. Richard & Knoppers Bartha Maria (eds.) Biotechnology IP & Ethics Lexis Nexis Canada Inc., Markham, Ontario p. 20–21

Goldstein Jorge A. & Elina Gold Elina (2002) Human Gene Patents. Academic Medicine 77:1316 (2002)

Gulliford Michael John (2004) Much Ado About Gene Patents: The Role of Forseeabiliity. Seton Hall Review 34: 720–721 (2004)

Holman Christopher M. (2007) Patent border wars: defining the boundary between scientific discoveries and patentable inventions. TRENDS in Biotechnology 25: 542 (2007)

Indian Patent Office, Annual Report (2007–08). http://ipindia.gov.in/cgpdtm/AnnualReport_English_2007-2008.pdf. Accessed 20 September 2012

Jameson Samantha A. (2009) A Comparison of the Patentability and Patent Scope of Biotechnological Inventions in the United States and the European Union. American Intellectual Property Law Association Quarterly Journal 35: 193–202

Jauhar Ameen & Narnaulia Swati (2010) Patenting Life the American, European and Indian Way. 15 Journal of Intellectual Property Rights 15: 60

Kankanala Kalyan C. (2007) Genetic Patent: Law and Strategy. Manupatra Information Solutions Pvt. Ltd., Noida, India

Kevles Daniel J. (2011) Patenting Life: A Historical Overview of Law, Interests, and Ethics. http://www.sba.oakland.edu/faculty/lauer/MIS641b/Readings/patenting%20life.pdf. Accessed 22 June 2011

Klein Roger D (2007) Gene Patents and Genetic Testing in the United States. Nature Biotechnology 25: 989

Lakshmikumaran Malthi (2007) Patenting of Genetic Inventions. Journal of Intellectual Property Rights 12: 51

Law Grace S. & Marles Jennifer A. (2004) Monsanto v. Schmeiser: Patent Protection for Genetically Modified Genes and Cells in Canada. Health Law Review 13: 44.

Lemley Mark, Things You Should Care About in the New Patent Statute http://ssrn.com/abstract=1929044 20 September 2011

Manual Patent Practice and Procedure, 2005

Stephen A. Merrill & Ann-Marie Mazza (eds) 2006a Reaping the Benefits of Genomic and Proteomic Research- Intellectual Property Rights, Innovation, and Public Health. The National Academies Press, Washington D.C 76.

Merrill Stephen A. & Mazza Ann-Marie (eds.) (2006b) Reaping the Benefits of Genomic and Proteomic Research- Intellectual Property Rights, Innovation, and Public Health. The National Academies Press, Washington D.C. p. 72

Mueller Janice M. (2008) Biotechnology Patenting in India: Will Bio-Generics Lead a "Sunrise Industry" to Bio-innovation? University of Missouri-Kansas City Law Review 75: 437–490; University of Pittsburgh Legal Studies Research Paper No. 2008-02. http://ssrn.com/abstract=1087131. Accessed 29 July 2012

Anita Nador, The Patenting of Biotechnology in Canada. http://www.samedanltd.com/magazine/12/issue/44/article/1245. Accessed 29 July 2012

Nese Bryan (2009) Bilski on Biotech: The Potential For Limiting The Negative Impact Of Gene Patents. California Western Law Review 46: 152

Nuffield Council of Bioethics (2002) The ethics of patenting DNA, 5.24. http://www.nuffieldbioethics.org/sites/default/files/The%20ethics%20of%20patenting%20DNA%20a%20discussion%20paper.pdf. Accessed 18 May 2011

Organisation for Economic Co-operation and Development (1999) Modern Biotechnology and the OECD http://www.oecd.org/SearchResult/0,3400,en_2649_201185_1_1_1_1_1,00.html. Accessed 15 May 2011

Perry Mark (2008) Pasteur to Monsanto: Approaches to Patenting Life in Canada. In: Gendreau, Ysolde (Ed.) An Emerging Intellectual Property Paradigm—Perspectives From Canada Ch. 4,72 Edward Elgar, Cheltenham, UK

Pila Justine (2003) Bound Futures: Patent Law and Modern Biotechnology 9.2 Boston University Journal of Science and Technology Law 9.2. http://www.bu.edu/law/central/jd/organizations/journals/scitech/volume92/pila.pdf. Accessed17 May 2011

Rai, Arti. K. (1999) Intellectual Property Rights in Biotechnology: Addressing New Technology. *Wake Forest Law Review* 34: 833

Rimmer Matthew (2008) Intellectual Property and Biotechnology: Biological Inventions. Edward Elgar Publishing, Inc., Massachsetts p. 24

Yelpaala Kojo (2000) Owning the Secret of Life: Biotechnology and Property Rights Revisited" 32 McGeorge Law Review 32: 11–219 Available via West Law. www.international.westlaw.com. Accessed 15 May 2011

Zahl Adrian (2004) Patenting of "Higher Life Forms in Canada. Biotechnology Law Report 23: 556

Zekos Georgios I. (2006) Nanotechnology and Biotechnology Patents. International Journal of Law and Information technology 14: 310. www.international.westlaw.com. Accessed 19 May 2008.

# Chapter 3
# Patentability of Biotechnology Under the International Patent Regime: Differentiation v. Harmonisation

International patent regime faces a great challenge to cope up with the new biotechnological advances. The international patent regime struggles to provide effective patent protection to biotechnology inventions (especially genetic inventions). The TRIPS Agreement sets minimum standards for the member nations to follow while granting patents; however, it leaves potential gaps and uncertainties as to the scope of numerous terms such as invention, microorganisms, microbiological processes and essentially biological processes. These gaps and uncertainties affect developing countries seriously given their relatively slow pace of scientific and technological development. The technology-neutral character of TRIPS does not allow special treatment to biotechnology inventions. International patent regime is the result of the efforts made by member nations to harmonise the patent laws and provide a uniform set of standards for the world. However, in the context of biotechnology patents, the divergence in patent practices among member nations makes it difficult to provide a uniform standard for the whole world. Moreover, there is a political divide between developed and developing countries as developed countries push for expanding the scope of patent eligible subject matter boundary by eliminating the exceptions from the text of TRIPS while developing countries are against this approach. In the biotechnological context, creating a single set of patenting guidelines for the entire world has proved very difficult to achieve given the controversy over issues such as patenting plants and animals. Both uniformity and diversity have potential and pitfalls, and the relevant question is to what extent inter-jurisdictional diversity and competition should be sacrificed to achieve global uniformity.

## 3.1 Territorial Nature of Patents

It is clear from the foregoing chapter that jurisdictional diversity exists among the USA, Canada, European Union (EU) and India regarding biotechnology patents. They have different patent approaches to the biotechnology inventions according to their relative social, economic and political conditions. This is in the line of territorial nature of patents as 'the institution of a patent regime is a matter for national sovereignty and the particular content and configuration of rules in such a regime is a matter for the laws of each nation state.'[1] Therefore, nations differ in the scope and coverage of patent protection to biotechnological inventions. Even in nations, which have similar patent laws, courts differ significantly in their interpretation of those laws. Under the territoriality principle, 'the same set of facts in a patent dispute can lead to conflicting judgements and irreconcilable outcomes when adjudicated in different countries.'[2] The territorial nature of patents created uncertainty regarding the grant of patents to the potential patent applicants and litigants. Further, it raised the cost of acquiring, protecting and enforcing patents in multiple nations. These deficiencies in the territoriality principle led to the internationalisation of patent system.

## 3.2 Internationalisation of Patent System: From Territorial to Global Patent Regime

The most relevant international agreements which led to the internationalisation of patent system are the Paris Convention on the Protection of Industrial Property (Paris Convention) and the Agreement on Trade Related Aspects of Intellectual Property Rights (hereinafter TRIPS Agreement). The Paris Convention was founded on the principle of national treatment, which ensures the consistent application of the patent laws of all signatories, but does not define their substance in any significant way.[3] The Paris Convention necessitates that the signatory member states of the World Trade Organisation (WTO) provide foreigners the same patent protections given to their own citizens.[4] It also prescribes minimum protective standards to be incorporated into domestic laws.[5]

Although the Paris Convention made a significant contribution in the internationalisation of the patent system, it had suffered from numerous deficiencies: Its main objective was restricted to eradicate discrimination rather than to harmonise different patent regimes; the inherent flexibility afforded by the national treatment principle under the convention accommodated consistently weak protection of patents, and the convention had also failed to specify minimum terms for the length of patents and finally, it lacked meaningful dispute settlement mechanisms.[6]

In order to correct the shortcomings of the Paris Convention, the TRIPS Agreement was adopted within the framework of the Uruguay Round of Multilateral Trade Negotiations. The TRIPS incorporated intellectual property rights (IPR) issues into the General Agreement on Tariffs and Trades (GATT). Developed countries advocated the inclusion of the intellectual property rights issues in the GATT

---

[1] Loughlan (1998).

[2] Bera (2009).

[3] Smith (2000).

[4] *Ibid.*

[5] *Ibid.*

[6] *Ibid.*

discussions by arguing that the inclusion would assist in liberalising the international trade; ensuring more effective enforcement of IPR; and facilitating negotiations regarding benefits and concessions across trade areas. Developing countries were initially reluctant to the incorporation of IPR provisions into GATT discussions; however, they were finally convinced about the said issues and adopted more free market policies. They also felt pressure from bilateral agreements.[7] Consequently, the Agreement on Trade-Related Aspects of Intellectual Property Rights (TRIPS) was concluded when the WTO Agreement replacing GATT was formally signed at Marrakesh on 15 April 1994.[8] TRIPS Agreement became effective from January 1995 with the formation of the WTO replacing GATT. It came into force from 1 January 1996.[9] The TRIPS Agreement has for the first time established international minimum standards before the member nations of the WTO regarding the grant of patent protection. The member nations are under obligation to implement these standards by bringing significant transformations in their respective patent laws.[10]

## 3.3 Patentability of Biotechnology Under TRIPS: Interpreting TRIPS in the Light of Biotechnology Inventions

Sec. 5 of the TRIPS Agreement deals with the patents, and Art. 27 of the TRIPS Agreement prescribes the requirements for patentability.
Art. 27.1 states:

> Subject to the provisions of paragraphs 2 and 3, patents shall be available for any inventions, whether products or processes, in all fields of technology, provided that they are new, involve an inventive step and are capable of industrial application. Subject to paragraph 4 of Article 65, paragraph 8 of Article 70 and paragraph 3 of this Article, patents shall be available and patent rights enjoyable without discrimination as to the place of invention, the field of technology and whether products are imported or locally produced.[11]

Article 27.1 sets significantly broad standard for subject matter of patent protection, prescribing that patent protection be made available for any inventions, whether products or processes, in all fields of technology, provided that they are new, involve an inventive step, and are capable of industrial application. Article 27.1 also emphasises that patents shall be available and patent rights enjoyable without discrimination as to the place of invention, field of technology involved subject to certain restrictions.[12] However, TRIPS does not provide a precise definition of the

---

[7] *Ibid.*
[8] Habiba (2009).
[9] *Id. at 65.*
[10] McManis (2003).
[11] Art. 27.1 of the TRIPS Agreement.
[12] McManis, *supra* note 10, at 83–84.

term, 'invention' and leaves it open for the individual nations to interpret. This leaves great uncertainty as to the precise scope of invention.

Highlighting the controversy attached to the interpretation of the term invention, Prof. Correa maintains that 'one of the most likely TRIPS controversies confronting the WTO Appellate Body will be over what constitutes an invention within the meaning of Art. 27.1, particularly, with respect to biological materials that pre-exist in nature.'[13] He further adds that one of the areas where the lack of a uniform concept of invention is most relevant relates to the distinction between patentable 'inventions' and unpatentable 'discoveries'.[14] According to him, a discovery is the mere recognition of what already exists; namely finding the casual relationships, properties, or phenomena that objectively exists in nature, while an invention is said to entail developing a solution to a problem by the application of technical means.[15]

### 3.3.1 Different Countries Interpret the Term 'Invention' Differently

Different countries approach the issue of patentability of a subject matter differently. For instance, the USA considers an isolated or purified form of a natural product patentable according to principles developed for chemical patents if it satisfies other patentability criteria viz. novelty, utility and non-obviousness. Similarly, European Patent Convention (EPC) allows patent for a substance found in nature if it may be characterised by its structure, the method of obtaining it, or by other criteria. A similar approach has been followed by Japan.[16]

In comparison to developed countries, developing countries, particularly Brazil have adopted a more restrictive approach regarding the patentable subject matter. Brazilian patent law does not allow patents on any invention claiming 'the whole or part of natural living beings and biological materials found in nature, or isolated there from, including the genome or germplasm of any natural living being, and any natural biological processes'.[17] Other developing countries adhere to a similar approach because they feel that despite having rich biological resources, they are in disadvantageous position as compared to developed countries, which possess the technology and resources to unveil and extract the value embodied in such materials.[18] The reference to place of invention in Art. 27.1 led to the USA to change its patent system, which was previously discriminatory against foreign inventions as the US courts did not accept foreign evidence of dates of invention. Art. 27.1 of the

---

[13] *Id.* at 88.

[14] *Id.* at 88–89; referring to Carlos M. Correa, Intellectual Property Rights, The WTO and Developing Countries; The TRIPS Agreement and Policy Options, 63 (2000).

[15] *Id.* at 89.

[16] *Ibid.*

[17] *Ibid. See* Brazilian Patent Law Article 10.1X, Law 9279, (1996).

[18] *Id.* at 89–90.

TRIPS is favourable to high technology corporations who oppose any discrimination on the basis of technology.[19]

### 3.3.2 Special Legislations for Different Technologies in Member Countries Violate Non-discrimination Provision Under TRIPS

*The European Biotechnology Directive Violates Article 27.1 of TRIPS* Since Art. 27.1 of TRIPS does not allow special treatment to be afforded to specific technologies, it can be argued that the EU constitutes a direct violation of art. 27.1 of TRIPS by providing special legislation in the form of EU Directive on the Legal Protection of Biotechnological Inventions 1998. The member states are under obligation to amend their patent laws in accordance with the directive to provide patent protection to biotechnological inventions. Art. 27.1 of TRIPS Agreement requires that member states enact laws that require only 'inventions' to be considered for patentability equally and consistently across 'all fields of technology'. By distinguishing 'biotechnological inventions' so that 'biological material' is presumed to be an 'invention', the requirement upon member states to amend their patent laws in accordance with the directive constitutes a direct violation of Art. 27.1 of TRIPS Agreement.[20]

In the case, *The Netherlands (supported by Italy and another) v. European Parliament and another (supported by the European Commission)*,[21] the validity of the directive was challenged.[22] However, the challenge was not directly concerned with the directive's conflict with Art. 27.1 of TRIPS. The challenge was made on the basis that the directive was removing the rights of member states to exclude from patentability 'plants and animals other than micro-organisms, and essentially biological processes for the production of plants or animals other than non-biological and microbiological processes' as provided by Art. 27(3)(b) of the TRIPS Agreement. Here, the contentious issue was whether the directive violates Art. 27.3(b) of the TRIPS Agreement.[23]

Responding to this contention, the court held that in narrowing the scope of excludable patentable subject matter, the directive was not inconsistent with TRIPS because the European Community was merely exercising an option provided by art. 27(3)(b) of TRIPS, and not therefore acting contrary to an obligation imposed.[24] The decision of the court that 'the Community legislative framework itself is not

---

[19] Dutfield (2009).

[20] Palombi (2003).

[21] *The Netherlands (supported by Italy and another) v European Parliament and another (supported by the European Commission)* [2002] All ER (EC) 97.

[22] Palombi, *supra* note 20, at 34.

[23] *Ibid.*

[24] *Id.* at 35.

illegal' did not explain the question why biotechnology should be given special treatment, when Art. 27.1 of TRIPS does not impose such a specific obligation.[25]

Therefore, it is argued that the patentability conditions must be applied equally. The special treatment to biotechnology raises concern about the appropriateness of the patent system as a vehicle of intellectual property protection for isolated genetic materials.[26] Professor Eisenberg rightly commented on the feasibility of the existing patent system in coping with the new technologies as 'the patent system was created for "a bricks and mortar world" which has inherent and logical limitations when transposed into the seemingly unlimited expansion of patentable subject matter.'[27] There requires a fresh approach regarding new technologies such as biotechnology by revising the patent system rather than patching the Biotechnology Directive to the biotechnology hole in the patent system.[28]

### 3.3.3 Exceptions Under the TRIPS Agreement

Arts. 27.2 and 27.3 of TRIPS create important subject-matter exceptions to the broad rule of Art. 27.1. Art. 27.2 states:

> Members may exclude from patentability inventions, the prevention within their territory of the commercial exploitation of which is necessary to protect *ordre public* or morality, including to protect human, animal or plant life or health or to avoid serious prejudice to the environment, provided that such exclusion is not made merely because the exploitation is prohibited by their law.[29]

Article 27.3 specifically deals with biotechnological inventions and important exclusions from patentability. Art. 27.3 reads as:

Members may also exclude from patentability:

a. diagnostic, therapeutic and surgical methods for the treatment of humans or animals;
b. plants and animals other than micro-organisms, and essentially biological processes for the production of plants or animals other than non-biological and microbiological processes. However, Members shall provide for the protection of plant varieties either by patents or by an effective *sui generis* system or by any combination thereof. The provisions of this subparagraph shall be reviewed 4 years after the date of entry into force of the WTO Agreement.[30]

It is clear from Art. 27.2 that the exception prescribed under it is subject to two important conditions; first, the exception applies only if a prohibition against the

---

[25] *Ibid.*

[26] *Ibid.*

[27] *Ibid.*; quoting Rebecca S. Eisenberg, 'Re-examining the role of patents in appropriating the value of DNA Sequences', 49 *Emory Law Journal* 783 (2000).

[28] *Ibid.*

[29] Art. 27.2 of the TRIPS Agreement.

[30] *Id.* Art. 27.3.

commercial exploitation of the invention is necessary to protect *ordre* public or morality; and second, the exception applies only if the exclusion from patentability will likewise contribute to the protection of *ordre* public or morality.[31] Article 27.3 allows WTO member countries to exclude from patentability two specific clauses of subject matters: (1) diagnostic, therapeutic and surgical methods for the treatment of humans or animals; and (2) plants and animals other than microorganisms, and essentially biological process for the production of plants or animals other than non-biological and microbiological process.

***Ordre* Public and Morality** Art 27.2 provides no definition of the terms *ordre* public and morality although references to human, animal or plant life or health or environment provide some context. The language of Art. 27.2 is nearly the same as that of the EPC, which shows the promotion made by the EU for its inclusion in the text of the TRIPS and the support provided by the developing countries.[32]

The implementation of these exceptions ensures that a WTO member may, in certain cases, refuse to grant a patent when it deems it necessary to protect higher public interests. The term '*ordre* public', derived from French law, is not an easy term to translate into English, and therefore, the original French term is used in TRIPS. It expresses concerns about matters threatening the social structures which tie a society together i.e. matters that threaten the structure of civil society as such.[33] Morality is the degree of conformity to conventional standards of moral conduct.[34] The concept of morality is a relative concept which depends upon the values prevailing in a society. These values differ from society to society and country to country and also change over time. The decision as to the patentability of a subject matter sometimes depends upon the moral judgement of that subject matter in that society.[35] Art 27 enunciates that protection of *ordre* public or morality includes the protection of 'human, animal or plant life or health or to avoid serious prejudice to the environment', thereby explicitly allowing for exceptions to patentability when any of these interests may be negatively affected by patent grants.[36]

The *ordre* public and morality exception under Art. 27.2 can only be applied if it is necessary to prevent the commercial exploitation of the invention. It is followed from this that the condition for the application of the exception would not be met if there is a need to prevent non-commercial uses of the invention (e.g. for scientific research). Here, the contentious issue is whether the exception can only be applied when there is an actual prohibition on the commercialisation of the invention, or when there is need to prevent it (even if still not done by the government concerned). In the light of Art. 27.2, the *ordre* public and morality exclusion would be

---

[31] McManis, *supra* note 10, at 83–84.

[32] Dutfield, *supra* note 19.

[33] Resource Book on TRIPS and Development 375 (Cambridge University Press, New York, 2005).

[34] Available at http://www.thefreedictionary.com/Moral+system (last visited on June 20, 2012).

[35] *Supra* note 33.

[36] *Id.* at 376.

justified when a member state demonstrates that it is necessary to prevent, by whatever means, the commercial exploitation of the invention. It would not be necessary for a member state to prove that under its national laws the commercialisation of the invention is actually prohibited.[37] Therefore, Art. 27.2 provides considerable flexibility for member countries to define situations in which the *ordre* public and morality exception would apply according to their respective public values.[38] The exception provided under Art. 27.3(a) reflects the acknowledged public interest in stimulating widespread dispersion of therapeutic innovations.[39]

### 3.3.4 Patenting of Life Forms Under the TRIPS Agreement: Internationalisation of Gene Patents

The US Supreme court's landmark decision made a great impact in expanding the scope of biotechnology inventions by holding that 'anything under the sun that is made by man' was patentable.[40] By making such interpretation, the court encompassed both foreseeable and unforeseeable subject matter including gene sequences. An imitation effect rippled from the USA to EU and other jurisdictions, prompting a series of legislative measures to patent living forms. Moreover, TRIPS internationalised biotechnological practices and enabled genetic engineering to yield important breakthroughs in the new millennium.[41]

**Patenting of Microorganisms and Microbiological Processes** Art 27.3(b) of TRIPS mandates that member states must provide patent protection to microorganisms and microbiological processes, however, it neither defines nor sets any parameter regarding the scope of the terms microorganisms and microbiological processes. Therefore, a distinct disconnect exists between the patentable subject matter in Article 27.3(b) and life forms that may be excluded from patent protection because the former has no commonly accepted definition in international patent law. Member countries, which have a well-established tradition of biotechnology patents such as the USA, Europe and Japan, differ in their interpretation of the patentability of subject matter. A patent system, therefore, may consider a plant cell a microorganism patentable despite the fact that it can grow into a tree. In such a situation, a patent on a cell could extend to trees even if one cannot patent a plant variety.[42]

There is a great uncertainty as to the scope of the term microorganism because even in scientific practice, the said term is inherently flawed as scientific classification continually evolves. In the medical literature, there is still no unanimously

---

[37] *Ibid.*
[38] Habiba, *supra* note 8, at 70.
[39] Smith, *supra* note 3.
[40] *Diamond v. Chakrabarty*, 447 U.S. 303, 309 (1980).
[41] Curci (2005).
[42] *Ibid.*

## 3.3 Patentability of Biotechnology Under TRIPS

accepted definition of microorganism, neither a certain boundary between microbiology and biology. Microorganisms have been defined broadly as unicellular structures with much reduced dimensions and not visible to the naked eye.[43] Another definition of microorganism is found in the decision of the Technical Board of Appeal of the EPO in the *Plant Genetic Systems* case: 'the term "microorganism" includes not only bacteria and yeasts, but also fungi, algae, protozoa and human, animal and plant cells, i.e. all generally unicellular organisms with dimensions beneath the limits of vision which can be propagated and manipulated in a laboratory. Plasmids and viruses are also considered to fall under this definition.'[44] The EU has decided to discontinue the use of the term 'microorganism'; instead, it has decided to use the term 'biological material', which means any material that contains genetic information and is capable of replicating itself or being reproduced in a biological system.[45]

There has been a political divide between developing and developed countries as to the extension of the concept of microorganism to include animal cells. Developing countries are against such extension, while many scholars in the industrialised countries advocate for maintaining the wording of Art. 27.3(b) of the TRIPS Agreement as such (i.e. without a precise definition) because microbiology is a fast-moving field of science and to provide a fixed and immutable definition of microorganism and microbiological process would reduce discretional powers of states at the implementation stage.[46]

Besides the term microorganism, it is also very difficult to define the term microbiological processes. Microbiological processes are generally thought to involve the use of microorganisms such as 'viruses, algae, bacteria and even cells or cell lines'. Since the definition of microorganism varies from country to country, therefore, it becomes difficult to define the term microbiological process too. Moreover, ambiguity remains as to how processes involving only microbiological steps are to be treated.[47]

**Essentially Biological Processes** Art 27.3(b) of the TRIPS Agreement provides that member states may also exclude from patentability 'essentially biological processes for the production of plants or animals'.[48] This exception has its roots in the European law and 'is generally thought to turn on the degree of technical intervention involved in creating the process'.[49] The greater the need for intervention, the less likely the process is to be classified as 'essentially biological' and the more

---

[43] Enrico Bonadio, "Biotechnology and patent law", available at http://www.dpsd.unimi.it/Italian_Intellectual_Property/archive/biotech.pdf. (last visited May 2, 2012).

[44] *Green Peace Ltd v. Plant Genetic System N.V.* (Case no. T 0356/93-334 dated 21-02-1995) Point 34.

[45] Sekar and Kandavel (2004).

[46] Bonadio, *supra* note 41; *See* the position taken by the European Community in the WTO document IP/C/W/383; *See* also Blakeney (2002).

[47] Smith, *supra* note 3.

[48] Art. 27.3(b) of TRIPS.

[49] Smith, *supra* note 3.

likely it is to be patentable.⁵⁰ In contrast, non-biological and microbiological processes related to the production of plants or animals are patentable under the text of Article 27.3(b).⁵¹ A non-biological process refers primarily to a therapeutic treatment of plants that is generally recognised as patentable in Europe.⁵²

**Meaning of Essentially Biological Processes** The correct meaning of the said article depends upon the interpretation of the words 'essentially biological'.⁵³ An interpretation provided by the Art. 2.2 of the EC Directive states that 'a process for the production of plants or animals is essentially biological if it consists entirely of natural phenomena such as crossing or selection.'⁵⁴ It can be inferred from this that according to the EC Directive genetic engineering processes and most modern biotech processes (which obviously do not consist entirely of natural phenomena) must be considered as patentable. Also, the EPO shares the above opinion.⁵⁵

For example, in the Plant Genetic Systems case the Technical Board of Appeal provided a definition of what does not constitute an essentially biological process: 'a process for the production of plants comprising at least one essential technical step, which cannot be carried out without human intervention and which has a decisive impact on the final result.'⁵⁶ Therefore, according to the EC Directive and the EPO case law, in order to consider the said inventions as patentable, it would be sufficient that only one segment of the production process is not biological.⁵⁷ However, the above view has been criticized by some commentators, who think that the said TRIPS provision should be interpreted in the opposite way.⁵⁸

**Plant Varieties** Art. 27.3(b) requires that 'members shall provide for the protection of plant varieties either by patents or by an effective sui generis system or by any combination thereof.'⁵⁹ This flexibility under this provision reflects the international divergence of views on protecting plant varieties as different countries allow different type of protection for plant varieties.⁶⁰ Along with patent protection and sui generis protection, Article 27.3(b) also provides an option before the members to combine patent systems with other forms of IPRs, further expanding the possible

---

⁵⁰ *Ibid.*; According to this notion, "classical breeding methods are not patentable," but genetic engineering methods are patentable.

⁵¹ Art. 27.3(b) of TRIPS Agreement.

⁵² Smith, *supra* note 3 (Non-biological processes would also include cultivation methods).

⁵³ *Ibid.*

⁵⁴ *See* Art. 2.2 of the EC Directive.

⁵⁵ Bonadio, *supra* note 43.

⁵⁶ *Supra* note 44 (See point 28 of the Decision).

⁵⁷ Bonadio, *supra* note 43.

⁵⁸ *Ibid.*

⁵⁹ Art. 27.3(b) of TRIPS Agreement.

⁶⁰ Smith, *supra* note 3. In Europe, plant varieties are protected under "breeder's rights" and may not be patented. In both Japan and the United States, however, plant varieties are patentable.

range and methods of protection.[61] Art. 27.3(b) contains ambiguous language, leading to conflicting interpretations.[62] For instance, it can be argued that an application relating a genetically engineered plant is bound to include plant varieties within its scope whether or not the word, 'variety' even appears in the specification. This ambiguity is problematic as in some jurisdictions plants can be patented but plant varieties cannot while in others neither can be patented.[63]

### 3.3.5 Article 27.3(b) of the TRIPS Agreement: A Temporary Compromise

The draft of Art. 27.3(b) of the TRIPS Agreement indicates a temporary compromise among the many competing interests in the protection of biotechnology, which is evidenced by the inclusion of an early revision date for these provisions (January 1999).[64] This article is the sole provision in the TRIPS Agreement subject to an early review due to controversial nature of special protection given to some specific inventions.[65] The framers of the said article anticipated a negotiated revision of the terms of Article 27.3(b) as the primary way of resolving this controversy.[66] The member countries seek modification and clarification of controversial terms via the WTO's administrative committees and dispute settlement procedures.[67] Members differ regarding the review of the said provision, while some members maintain that the review should be an examination of the extent to which the current provisions have been implemented,[68] others advocate for a more substantive process that might encompass changing the text of the article.[69] Developed countries insist that Art. 27.3 merely speaks for a review of implementation while developing countries maintain that 'the review should include the possibility of revising the text not

---

[61] *Ibid.* (For example, a system combining patents with other forms of IPR protections might allow for incorporation of "farmer's rights" which recognize and compensate for the "ancestral contributions" of traditional farmers in developing new plant varieties).

[62] *Ibid.* (The three-sentence article contains nine terms whose meanings are open to debate: "plants, animals, micro-organisms, essentially biological processes, non-biological, microbiological, plant varieties, effective and sui generis system").

[63] Dutfield, *supra* note 19 at 250.

[64] Smith, *supra* note 3; Art. 27.3(b) of TRIPS Agreement. (The provisions of this subparagraph shall be reviewed 4 years after the date of entry into the force of the WTO Agreement. The WTO Agreement entered into force on January 1, 1995.)

[65] *Ibid.*

[66] *Ibid.*

[67] *Ibid.*

[68] This is the approach advocated by most developed nations. Although these countries would eventually like to see many of the vague terms of Article 27.3(b) defined or deleted, they are concerned that any immediate attempt to change the terms will lead to a weakening of IPR provisions. *Ibid.*

[69] *Ibid.* (This is the viewpoint espoused by many developing nations unhappy with any delineation of plants and animals as patentable materials).

so much to strengthen but rather to loosen Art. 27's requirements with respect to patentable subject matter, and in any event to make this provision of TRIPS more sensitive to the needs and interests of the developing world.'[70]

Developing countries propose that Art. 27.3(b) should be revised to ensure that naturally occurring materials, including genes are not patentable. They insist that it should also recognise the adequate protection of traditional knowledge of local and indigenous communities.[71] Developing countries seek that the exception for plants and animals should be maintained in the TRIPS Agreement. They also seek that they should have the flexibility to develop sui generis regimes on plant varieties, suited to the seed supply systems of the countries concerned.[72] Developing countries also proposed in the TRIPS Council and before the WIPO, 'to establish international patent standards requiring members not to grant or to cancel *ex officio* or upon request any patent or other intellectual property rights on any biological materials obtained (1) from collections held in international germplasm banks and other depositaries where such materials are publicly available; or (2) without the prior consent of the country of origin and/or indigenous or local communities providing the materials.'[73]

Developing countries further demand compliance with obligations contained in the Convention of Biological Diversity (CBD), particularly, to share the benefits with the country of origin of any patented biological material.[74] This concern was recognised in the Fourth Ministerial Conference of the WTO in Doha, where, the Doha Ministerial Declaration instructed the TRIPS Council to examine, inter alia, 'the relationship between the TRIPS Agreement and the convention on Biological Diversity, the protection of traditional knowledge and folklore, and other relevant new developments raised by members pursuant to Art. 71.1 [which authorizes the TRIPS Council to undertake reviews in the light of any new developments which might warrant modification or amendment of this agreement].'[75]

However, under the current political climate where developed and developing countries differ significantly, it is very unlikely that exceptions from patentability permitted by Art. 27.3(b) will soon be eliminated. Simultaneously, it is also not likely that in the upcoming round of multilateral trade negotiations at least some of the foregoing demands of the developing world will be met.[76] As regards to human genome research, the immediate implications are not quite profound, however,

---

[70] McManis, *supra* note 10 at 93–94.

[71] *Ibid.*

[72] *Ibid.*

[73] *Ibid.*

[74] *Ibid.*

[75] *Ibid*; *See* Doha WTO Ministerial 2001: Ministerial Declaration, WT/MIN(01)/DEC/1, Nov. 20, 2001, adopted Nov. 14, 2001, para. 17 and 19 [hereinafter Doha Declaration] para. 19; *See* also para. 17 (stressing the importance of implementing and interpreting the TRIPS Agreement "in a manner supportive of public health, by promoting both access to existing medicines and research and development into new medicines," in connection with which the Doha Ministerial issued a separate, and more detailed declaration.

[76] *Id.* at 93–94.

3.3 Patentability of Biotechnology Under TRIPS                                    123

the international prospects for patent protection on upstream, microbiological or sub-microbiological genetic research look a good deal, more promising in the post-TRIPS world than the prospects for patent protection on downstream macrobiological commercial products.[77]

### 3.3.5.1 GATT Negotiation

The debate regarding the patentability of biotechnology products and processes has not been confined to North-South debate but has increasingly become a North-North debate.[78] The negotiating proposals of the USA, Japan, the Nordic countries and Switzerland initially advocated for broad patent coverage without exclusion for plants and animals. However, lately, most member countries, industrialised as well as developing, had opposed the patentability of plants and living organisms.[79] Consequently, the EPC[80] has become the obvious model for including certain exceptions under Art. 27.2 and 27.3. Article 52(4) of the EPC maintains that methods of treatment of the human or animal body shall not be regarded as inventions.[81] Art. 53 speaks about *ordre* public exceptions and specifies that patents shall not be granted for plants or animal varieties or essentially biological processes for the production of plants or animals, other than microbiological processes or the products thereof.[82] The USA asserted the fact during negotiations that there was no international consensus as to whether genetically engineered life forms should be patentable.[83] The EU joined the USA and Switzerland while showing concern over widespread exclusions for food, chemical and pharmaceutical product. These countries had also

---

[77] *Ibid.* (If isolated genetic sequences as such or isolated genetic sequences as inserted in an organism in which they do not appear in nature are indeed patentable subject matter, it will make little practical difference whether the larger macroorganisms into which they are inserted are or are not themselves patentable, as their genetic sequences, in any event, cannot be 'made used, offered for sale, sold, or imported without the consent of the patent owner).

[78] *Id.* at 82.

[79] *Ibid.*

[80] European Patent Convention, Oct. 5, 1973, as amended Dec. 21, 1978.

[81] McManis, *supra* note 10, at 82.

[82] *Id.*at 82–83; Compare Arts. 52(4) and 53 of the EPC with Art. 27.2 and 27.3. The text of EPC Articles 52(4) and 53 are as follows: "Article 52: Patentable Inventions (4) Methods for treatment of the human or animal body by surgery or therapy and diagnostic methods practiced on the human or animal body shall not be regarded as inventions which are susceptible of industrial application within the meaning of paragraph 1. This provision shall not apply to products, in particular substances or compositions, for use in any of these methods. Article 53: Exceptions to Patentability European patents shall not be granted in respect of: (a) inventions the publication or exploitation of which would be contrary to "*ordre* public" or morality, provided that the exploitation shall not be deemed to be so contrary merely because it is prohibited by law or regulation in some or all of the Contracting States; (b) plant or animal varieties or essentially biological processes for the production of plants or animals; this provision does not apply to microbiological processes or the products thereof."

[83] *Id.* at 83.

expressed their concern over the limited recognition of sui generis protection of plant breeder's right.[84] Developing countries, on the other hand, raised their concern that they might be prevented from securing access to modern technology due to overprotection of IPR. Developed and developing countries reached to a compromise, where in return for increased developing country access to the agricultural and textile markets of the industrialised world, developing countries reluctantly agreed to Art. 27 of TRIPS, as tempered by the transitional patent provisions contained in Art. 65.4, which in turn qualified by the 'mailbox' and 'exclusive marketing rights' provisions of Art. 70.[85]

### 3.3.5.2 Review of Art. 27.3(b) of the TRIPS Agreement

A possible solution to this problem is seen in summarily executing a review, postponing any significant substantive or procedural changes until more information is available.[86] Such an approach would maintain the tentative compromise reached on IPR issues relating to biotechnology and life forms.[87] However, opponents of this approach express their concern regarding the possibility that members may attempt to bypass WTO amendment procedures and define the vague terms of Article 27.3(b) through other means.[88]

One such possibility is enshrined in Art. 71.1, which empowers the TRIPS Council to 'undertake reviews in the light of any relevant new developments which might warrant modification or amendment of this Agreement.'[89] Some commentators maintain that this broad language gives the TRIPS Council the authority to issue statements of interpretation unreviewable by the Ministerial Conference so long as they only modify existing TRIPS terms.[90] However, whether the TRIPS Council has its power under Article 71 is a debatable issue.[91] It is also a debatable issue, whether the council would exercise that power if it were available.[92]

---

[84] *Ibid.*

[85] *Ibid.*

[86] Smith, *supra* note 3.

[87] *Ibid.*

[88] *Ibid.*

[89] *Ibid.*

[90] *Ibid.*

[91] *Ibid*; Article 71.2 of the TRIPs Agreement outlines the limited situation in which modification actions may be taken by force of recommendation from the TRIPs Council, stating that:
[a]mendments merely serving the purpose of adjusting to higher levels of protection of intellectual property rights achieved, and in force, in other multilateral agreements and accepted under those agreements by all Members of the WTO may be referred to the Ministerial Conference for action in accordance with paragraph 6 of Article X of the WTO Agreement on the basis of a consensus proposal from the Council for TRIPs.
It is unlikely that a more general ability to circumvent the amendment procedure was intended in Article 71.1.

[92] *Ibid.*;It is noteworthy, however, that the potential bypass of the consensus-based decision making of the Ministerial Conference was suggested specifically in relation to revision of Article 27.3(b).

The WTO Dispute Settlement Body is another possible way to bypass the Ministerial Conference. The WTO's Dispute Settlement Body's panels issue decisions, binding on the parties involved and set precedents for other members in future disputes.[93] These panels can be utilised to define nebulous terms and effectively limit the ability of member countries to determine the precise scope of IPR holders under domestic laws.[94] However, the main flaw attached with clarifying terms through case-by-case decision making is that it will produce short-sighted decisions with unfortunate local and international effects.[95]

No complaints have been filed yet regarding the terms of Art. 27.3(b)[96] and no attempts have been made to abuse the DSU process by circumventing the negotiating bodies of the WTO. Despite the lack of direct legal action, the threat of such action may still affect domestic policy decisions regarding interpretation of the terms of the article.[97] Attempts at substantive revision of Article 27.3(b) will pose additional challenges arising from the WTO structure.[98]

### 3.3.6 Other Patent Eligibility Criteria Under TRIPS

**Novelty, Inventive Step, Industrial Application, Written Description and Disclosure** The eligibility standards for the novelty and inventive step ensure free competition.[99] Industrial countries which make huge investments in the R&D may favour permissive novelty standards and low standards on the inventive step. However, despite an incremental effect in few sectors, this policy may lead to the growing number of patents that protect trivial developments.[100] It is a proven fact that that a higher innovative step requirement can increase the value of patents, because patents issued under this rule are stronger and less vulnerable to challenge by competitors.[101]

---

[93] *Ibid.*

[94] *Ibid.;* Article 27.3(b) is not the only provision of the TRIPs Agreement susceptible to interpretation and expansion through the DSU mechanism. For example, Article 41.5 gives member states the right not to augment their judicial and administrative systems in order to provide for TRIPs-required IPR protection. Many developing countries have rudimentary infrastructures and limited resources to administer the requisite IPR framework. Dispute panels are likely to be faced with the challenge of distinguishing between will full non-compliance with the obligations of the agreement and genuine lack of capacity to do so.

[95] *Ibid.*

[96] *Ibid.*

[97] *Ibid.*; Governments may refrain from following reasonable and appropriate interpretations of the Article's terms out of fear of opening themselves to costly legal challenges.

[98] *Ibid.*

[99] Correa (2001–2002).

[100] *Ibid.*

[101] *Ibid.*

Developing countries advocate for setting higher standards of novelty and inventive step in order to ensure competition without violating international minimum standards.[102] The World Bank echoes the same view, according to which, 'countries could set high standards for the inventive step, thereby preventing routine discoveries from being patented. Regarding patent scope, it is sensible to exercise strict claims and discourage multiple claims in patent applications.'[103]

**Deposition of Biological Material Mandated Under TRIPS** Article 29 of the TRIPS Agreement states that:

> Members shall require that an applicant for a patent shall disclose the invention in a manner sufficiently clear and complete for the invention to be carried out by a person skilled in the art and may require the applicant to indicate the best mode for carrying out the invention known to the inventor at the filing date or, where priority is claimed, at the priority date of the application.[104]

**Sufficient Disclosure and Best Mode** For all practical purposes, the same result is accomplished by the corresponding provision of Patent Cooperation Treaty (PCT), which includes describing the nucleotide sequences, deposition of microorganism to supplement the written description. The provisions as mentioned above also exist in the patent laws of almost all countries for the reason that when term of protection (patent) is over, the public should be able to take benefit of the invention. It is very difficult to describe the invention relating to biotechnology by written description as it involves the use of living material such as microorganism. Budapest Treaty provides facility to deposit the microorganism in any of the International Depository Authority (IDA) recognised by WIPO to supplement the written description to avoid deposition of such microorganism in each country where the applicant applies for grant of patent.[105]

Only few countries have a culture collection with IDA status (20 countries as of 4 April 2003).[106] It is, therefore, required to maximise efforts to increase the number of IDAs. It is equally important that IDAs should demarcate naturally isolated

---

[102] *Ibid.*; The degree of flexibility granted to countries to determine the level of inventive step requirement is illustrated by the history of the European patent system:... There is no reasonable exact measure for inventive step. It was known that the Dutch Patent Office has used such a high degree [of inventive step] that the percentage of grants was relatively low there. This was also known by the first President of the European Patent Office, Mr. Van Benthem, who was earlier the President of the Dutch Patent Office. Because of this experience, Mr. Van Benthem convened, before starting work at the European Patent Office, a conference of practitioners who submitted their views whether they preferred "a high, middle or lower degree of inventive step for the examination of European patent applications." *See* Heinz Bardehle, Regional Approaches: European Patent System, presentation to the Conference on The International Patent System, WIPO, Geneva, Mar. 25–27, 2002.

[103] *Ibid.*; *See* World bank, global economic prospects and the developing countries 2002, at 143 (2001).

[104] Art. 29 of TRIPS Agreement.

[105] Habiba, *supra* note 8, at 73.

[106] Sekar & Kandavel, *supra* note 45 at 214.

microorganisms from genetically modified microorganisms to preserve the natural gene pool. Another significant challenge is how to preserve the natural gene pool without the accumulation of spontaneous mutations and thereby to maintain the originality of the strains deposited.[107] In this regard, efforts should be made to promote the preservation of the natural gene pool.

## 3.4 Feasibility of a Uniform Global Patent System: Differentiation v. Harmonisation

As regards to patentability of biotechnology inventions, the issue of harmonisation of patent laws and formation of a uniform global patent system has gained much importance. The issue of harmonisation stands in conflict with the territorial nature of patents, which implies differentiation among different jurisdictions. Here, the pertinent questions are: What the potentials and pitfalls of harmonisation of patent systems are; whether harmonisation of patent laws to provide a unified patent system is justified given the territorial nature of patent laws; to what extent harmonisation of patent laws should be permitted against the flexibilities provided under international patent treaties such as TRIPS; what are the impediments in the harmonisation process; and finally, how signatory nations to WTO respond to the harmonisation process?

The process of global harmonisation of patent laws began in the twentieth century when international agreements on patents moved in the direction of establishing global minimum protective standards.[108] These standards had been derived primarily from the existing laws and philosophies of developed nations, leaving very little space for the divergent cultures and interests of less developed nations.[109] The remarkable breakthrough in this process was the creation of the WTO and the signing of the TRIPS Agreement. The TRIPS Agreement, for the first time, prescribed minimum standards for patentability before the member nations along with significant enforcement mechanisms, which revolutionised the international intellectual property law.[110]

Apart from TRIPS, various other measures have been taken thus far in the form of patent law treaty, and draft substantive patent law treaty (SPLT), to bring uniformity among the patent laws of member nations. However, the goal is not achieved yet as the harmonisation process has been facing numerous challenges and obstacles. One of the main obstacles in harmonisation process has been finding a common ground between different interests affected by the uniform patent standards.[111] These differences are noticeable in the area of protection of biotechnology innova-

---

[107] *Id.* at 214–215.
[108] Smith, *supra* note 3.
[109] *Ibid.*
[110] *Ibid.*
[111] *Ibid.*

tions, where patenting of life forms has significant social, moral, economic, cultural and religious implications.[112] This creates great hurdle in achieving a mutually agreed common ground between different nations regarding the appropriate protection of biotechnology inventions. Although a tenuous compromise on these issues was reached with the creation of Article 27.3(b) of the TRIPS Agreement, however, the scheduled review of the nebulous terms of this compromise may produce conflict over the specific standards set, as well as the processes used to establish them. Therefore, the situation warrants proceeding with caution.

The TRIPS Agreement has a great impact on the domestic patent laws of all WTO member states. Art. 27.1 embodies principle of equal treatment for all technologies,[113] which obligates member states to accommodate biotechnology inventions.[114] Therefore, the TRIPS Agreement compels member states to provide patent protection to biotechnology inventions. The TRIPS Agreement sets only minimum standards for patentability and disclosure, therefore a considerable variability exists between member states as to the definition of patent eligible inventions and adoption of optional exclusions from patentability of certain kinds of inventions.[115]

TRIPS sets a floor for patent protection but leaves the ceiling open. Therefore, developed nations with strong intellectual property laws insist their trading partners to increase intellectual property protection as part of a bilateral trade agreement.[116] For example, the USA and EU often require that the bilateral partner amend its patent laws to include a broader definition of eligible subject matter than the minimum set out under TRIPS.[117] More often, these so called TRIPS-plus agreements necessitate protection for biotechnology or prohibit exclusions on patenting plants and animals.[118] The bilateral free trade agreement between Jordon and the USA reflects such trend, where the said agreement limits exclusions to patentable subject matter only to that which is necessary to protect '*ordre* public or morality' and requires that 'diagnostic, therapeutic and surgical methods for the treatment of humans or animals be patentable'.[119]

---

[112] *Ibid.*

[113] Art. 27(1) of the TRIPS Agreement provides "patents shall be available…without discrimination as to…the field of technology"

[114] Keim (2007).

[115] *Ibid.*

[116] *Ibid.*

[117] *Ibid.*

[118] *Ibid.*; *See* Grain, *"TRIPS-plus" through the back door: How bilateral treaties impose much stronger rules for IPRs on life than the WTO* (2001) (available at http://www.grain.org/briefings/?id=6, last visited May 8, 2012).

[119] *Ibid.*; *See* The US-Jordan FTA, Arts. 18(a), 18(b).

## 3.4.1 Draft Substantive Patent Law Treaty

These inconsistencies and uncertainties regarding the patentability requirements led to the drafting of Substantive Patent Law Treaty (SPLT). Major patenting nations voiced for substantive patent harmonisation and the International Bureau of the World Intellectual Property Organization composed a draft SPLT in 2001. However, the continuous disagreement between the member states has put further discussion about the SPLT on indefinite hold.[120] The controversy over issues such as patentability of plants and other life forms made it difficult to provide a single set of patenting guidelines for the entire world. Moreover, developing countries push themselves back from harmonisation as they feel that TRIPS is already overprotective of intellectual property.[121]

## 3.4.2 Differentiation vis-a-vis Harmonisation

Harmonisation process should have a cautious approach because of numerous factors including challenges posed by new technologies. The feasibility, costs and benefits of a further harmonisation should be adjudged from an economic as well as legal perspective. However, there is dearth of economic studies on the said topic.[122] Further, there is no significant evidence in the favour of the argument that patents produce development and the member states should adopt substantially the same standards of patent protection, irrespective of their level of development.[123]

A recent report by the World Bank reflects that the patent system imposes high costs for developing countries because of numerous factors viz. administrative costs, high costs involved in key technological inputs, high prices of medicines etc., while 'long term benefits seem uncertain and costly to achieve in many nations, particularly for the poorest countries.'[124] The said report concludes that 'one size does not fit all', and that countries should be left with the flexibility to adapt the levels of intellectual property protection as their economies grow. It further argues that 'it should be recognized that developing countries need to have lower and more flexible IPR standards than do their developed counterparts. TRIPS provides such flexibility in many areas and it becomes pertinent that developing countries should

---

[120] *Ibid.*; *See* WIPO Magazine, *Patent Law Harmonization: What Happened?*, (2006) (available at http://www.wipo.int/wipo_magazine/en/2006/03/article_0007.html, last visited May 8, 2012).

[121] *Ibid.*; *See 2006 Patent* cooperation treaty conference: transcript of proceedings, 32 Wm. Mitchell L. Rev. 1603, 1655–1657 (2006).

[122] Correa, *supra* note 99.

[123] *Ibid.*

[124] *Ibid.*; *See* World bank, global economic prospects and the developing countries 2002, at 129 (2001).

be afforded the opportunity to operate at the lower limits if it is in their development interests to do so.'[125]

Harmonisation of patent laws is significant and challenging in the field of new technologies such as biotechnology. Despite the fact that that the difference as to the patenting of biological material has been significantly narrowed down by the Trilateral Agreement between the USA, EU and Japan, divergences still exist regarding the treatment of biotechnology inventions.[126] There is much divergence in the patent approaches of developed and developing countries as few of the latter have explicitly excluded the patentability of existing biological materials, unless they are genetically altered.[127] Under such patent scheme, the patent law excludes certain biotechnology-based products from patentability while allowing patents for the process used to obtain the biotechnology-based product. Furthermore, few of the developing countries consider the patentability of any life form as essentially contrary to basic ethical values, and wish to remain free from such protection.[128]

### 3.4.3 Merits and Demerits of Uniform Patent Law

Uniformity of law provides potential benefits for member countries; it simplifies the law, makes it easier to learn and describe, and reduces administrative costs. However, it has certain demerits also; it makes the law unresponsive to local variations, eliminates inter-jurisdictional competition and decreases the possibilities for legal experimentation.[129] International patent community has given utmost importance to the harmonisation of patent laws, creating uniform patent law on a global scale through the diversity of the existing systems. The most significant manifestation of this trend is the TRIPS Agreement, which necessitates that the patent laws of the signatory nations must be in conformity with a uniform framework of international standards.[130] Harmonisation of patent laws provides potential benefits not only to the world economy but to the individual nations as well.[131] Harmonisation of patent laws brings certainty, simplifies patent laws and combats the patent costs, which

---

[125] *Ibid.*; *See* World bank, global economic prospects and the developing countries 2002, at 147 (2001).

[126] *Ibid.*

[127] *Ibid.* (Thus, the Brazilian patent law (1996), stipulates that no patents shall be granted with respect to living beings or "biological materials found in nature," even if isolated, including the "genome or germplasm" of any living being).

[128] *Ibid.*; *See*, e.g., The proposal for review of article 27.3(b) of the TRIPS Agreement submitted by Kenya on behalf of the African countries, WT/GC/W/302, Aug. 6, 1999.

[129] Duffy (2002).

[130] *Ibid.*

[131] Smith, *supra* note 3.The primary benefit of harmonization is the encouragement of research in beneficial areas and the stimulation of the marketing and distribution of the fruits of those labors worldwide. In the short term, these benefits will accrue most directly to developed countries. Id. In the long run, however, developing countries are also expected to reap these benefits and profit from an attendant increase in the transfer of technology.

promotes competition and enhances world economic growth.[132] Harmonisation of patent laws in the post-TRIPS world continues to be a shibboleth in patent circles, and diversity a flaw to be remedied.[133]

There is no doubt that harmonisation of patent laws provides certain benefits, however, it also entails certain disadvantages as it would preclude inter-jurisdictional competition and experimentation in patent law, among other things. Here, 'the relevant policy question is to what extent inter-jurisdictional diversity and competition should be sacrificed to achieve global uniformity.'[134] This question is significant for understanding proper limitations of the measures already taken towards global harmonisation of patent law, especially the TRIPS Agreement. If jurisdictional diversity holds some merit, it becomes pertinent that the provisions under TRIPS permitting diversity and flexibility should be interpreted broadly.[135]

TRIPS prescribes broadly similar patent laws for all the member nations but does not delve much into the details of systems in part because implications of those details are not known exactly. Therefore, uncertainty remains as to which of the patent systems produces the better incentive to invent. In such a situation, TRIPS rightly leaves those matters open for each country's choice.[136] John Duffy summarises the debate regarding the constant struggle between diversity and uniformity of patent laws as:

> ... [i]n reforming current law, we should resist the Sirens' song of complete uniformity. A consolidation of existing patent systems into a single monolith would impoverish the field; it would be mass extinction of legal species. Diversity has its own worth; it permits competition and breeds innovation. These virtues should be evident to the patent community, for they are dear not only to the goals of the patent law, but also to its history. Patent law of the twenty-first century would be enriched if national and international policymakers learn to balance the values of harmony with those of cacophony.[137]

### 3.4.4 Relevance of the Existing International Patent Regime in the Present Technological Age

Patent laws of most countries embody premises and concepts that were influenced and shaped by the Industrial Revolution. These laws are ill-suited in the current information driven age. In the present age, there is abundance of inventions that spring from new technologies such as biotechnology, information technology and robotics.[138] Reichman and Dreyfuss have succinctly summarised the problem with the existing international intellectual property regime:

---

[132] Smith, *supra* note 3.
[133] *Ibid.*
[134] *Ibid.*
[135] *Ibid.*
[136] *Ibid.*
[137] *Ibid.*
[138] Bera, *supra* note 2.

[T]he worldwide intellectual property system has entered a brave new scientific epoch, in which experts have only tentative, divergent ideas about how best to treat a daunting array of emerging new technologies. The existing system has become increasingly dysfunctional because it operates with a set of rudimentary working hypothesis that have not kept pace with technical change.[139]

## 3.4.5 Tentative Harmonisation Efforts

TRIPS Agreement was the most notable step in the direction of harmonisation.[140] Article 1.1 gives member states freedom to 'to determine the appropriate method of implementing the provisions of this Agreement within their own legal system and practice'.[141] Moreover, a November 2005 decision of the council for TRIPS allowed least developed country members to postpone implementation of many TRIPS obligations until 2013.[142] Countries have the option to refuse patents on diagnostic, surgical and therapeutic methods.[143] They can also refuse patents on inventions that are required to protect *ordre* public, morality and human health.[144]

*Obstacles in Harmonisation* The experience with the TRIPS Agreement suggests that SPLT will need to cross many hurdles. Member states would have to found common ground regarding several issues: To agree upon what is 'patentable subject matter'; to settle upon common definitions of and articulated standards for terms such as 'novelty', 'non-obviousness' (or 'inventive step'), 'useful invention', 'doctrine of equivalents', 'infringement', etc.[145] The infrastructure capable of interpreting and amending the 'harmonized' law must be in place well in advance before the implementation of full-scale harmonisation occurs.[146]

*SPLT Is Futuristic* SPLT is futuristic as it has to cross many hurdles to get success.[147] Brazil had been opposing the SPLT and critical to the political orientation of WIPO. Subsequently, in 2004, the governments of Brazil and Argentina moved a proposal that WIPO adopt a 'Development Agenda'.[148] This led to the formation of a group of 14 developing countries called Friends of Development, which produced a follow-up submission to WIPO. This latter document elaborated on the first one and responded to some developed countries, especially the USA, that were opposed

---

[139] Reichman and Dreyfuss (2007).
[140] Bera, *supra* note 2.
[141] Art. 1.1 of the TRIPS Agreement.
[142] Bera, *supra* note 2.
[143] *Ibid.*; *See* Art. 27(3)(a) of the TRIPS Agreement.
[144] *Ibid.*; *See* Art. 27(2) of the TRIPS Agreement.
[145] Bera, *supra* note 2 at 456.
[146] *Ibid.*
[147] *Ibid.*
[148] Dutfield, *supra* note 19 at 273.

to any comprehensive initiative to explicitly incorporate development concerns into the mandate and activities of WIPO.[149]

The opponents of the substantive patent law harmonisation argue that the strength of patent protection should be commensurate with a country's level of development.[150] They further argue that such harmonisation process would have effectively prevented developing countries from designing patent law as an instrument of economic development.[151]

**World Trade Organisation (WTO)** Developing countries express their concern against the raising of intellectual property standards and proposed that TRIPS be revised in order to circumscribe certain rights, to maintain or even expand the exceptions, and to create new intellectual property frameworks.[152] Developed countries as well as life sciences corporations seek that patent standards be raised. There are few developing countries, which are in favour of maintaining the same text of the TRIPS Agreement; however, there are others, which support modification in the existing patentability standards under TRIPS.[153] In the due course of time, developed countries have softened their stance and shift their focus for the time being from raising the patentability standards further to implementing the existing standards. This shift was an outcome of developing countries' counter proposals and more proactive stance.[154] However, the USA and EU are changing their strategies to put pressure on developing countries to raise intellectual property standards beyond those required by TRIPS. They are doing so by sidestepping WTO and using different means such as threats to remove trade concessions, and carrots like bilateral and regional free trade agreements.[155]

These agreements compel developing country partners to undertake such commitments as: 'to introduce TRIPS standards before the expiry of the transitional periods, to provide higher standards of protection than TRIPS requires, and to extend the subject matter of intellectual property protection in ways that TRIPS may not necessarily require.'[156] In this situation, developing countries are in a disadvantageous position and lack any clear idea as to how biotechnology can benefit their economies and improve the lives of their citizens. They also lack the capacity to design an intellectual property system to promote welfare enhancing biotechnological innovation. Many of the developing countries have no biotechnology industry and they remain doubtful that such businesses will be grown just because life forms and microbiological processes can be patented.[157]

---

[149] *Ibid.*
[150] *Ibid.*
[151] *Id.* at 274.
[152] *Ibid.*
[153] *Id.* at 285.
[154] *Id.* at 286.
[155] *Ibid.*
[156] *Ibid.*
[157] *Ibid.*

However, the current trend suggests that the patent systems continue to evolve and life industries continue to be a driving force for the change. In such a situation, it is very unlikely that there will be any rolling back of the proprietary trend evident in the past few decades. Opponents may only succeed in delaying the changes or reducing its extent but will not be able to prevent them. Business and patent community have been given the interpretive custody of the patent system, as a result of which we have an unbalanced patent system. There is a great need for democratisation of intellectual property process at the international level, ensuring full public participation.[158]

Controversy also exists regarding the morality and public order exclusion under Art. 27.2 and animals and plants under Art. 27.3 of the TRIPS Agreement. The USA argues that there should be no exclusions to patentability in the SPLT.[159] However, Europe and the developing countries argue in the favour of maintaining the said exclusions in the SPLT. As things stand now, countries which sign the treaty will not be allowed to make any further demands on patent applicants than those found in the treaty. This led to the confrontation between industrialised and developing countries. Brazil, the Dominican Republic and Peru, among others, are arguing that the disclosure of country of origin of genetic materials, and proof of prior informed consent in their acquisition, must be enforced.[160]

## 3.5 Implications of Setting up a Uniform World Patent System

The setting up of a uniform world patent system would have great implications for the member countries that use patent as a tool for national development strategies. It would also overtake the TRIPS both in form and in substance.[161] The negotiation of the SPLT is facing a contentious debate between the USA and Europe. Europe along with Japan is thus far defending the status quo of TRIPS, with Japan following its line. Among the developing countries, Brazil is the most vocal on the disclosure issue. It is a debatable question whether the SPLT and the TRIPS Agreement can coexist or will conflict.[162] The conflicting interests of the parties involved will make it a very complicated and torturous process. However, the current development suggests that things are going steadily in the direction of a unified world patent system along with various impediments and frictions.[163]

---

[158] *Id.* at 287.

[159] GRAIN (2002), "WIPO moves toward "world" patent system", available at http://www.grain.org/system/old/docs/wipo-patent-2002-en.pdf (last visited on May 10, 2012).

[160] *Ibid.*

[161] *Ibid.*

[162] *Ibid.*

[163] *Ibid.*

# References

Bera Rajendra K. (2009) Harmonization of patent laws. Current Science 96: 457

Blakeney M. (2002) Access to Genetic Resources, Gene-based Inventions and Agriculture. www.iprcommission.org. Accessed 4 May 2012

Bonadio Enrico (2009) Biotechnology and Patent Law http://www.dpsd.unimi.it/Italian_Intellectual_Property/archive/biotech.pdf. Accessed 2 May 2012

Correa Carlos M. (2001–2002) Internationalisation of the Patent System and New Technologies. Wisconsin International Law Journal 20: 523 (www.international.westlaw.com. Accessed 5 October 2011)

Jonathan Curci (2005) The new challenges to the international patentability of biotechnology: legal relations between the WTO treaty on Trade-Related Aspects of Intellectual Property Rights and the Convention on Biological Diversity. International Law & Management Review 2:1. Available via West Law. www.international.westlaw.com. Accessed 15 May 2008

Duffy John F. (2002) Harmony and diversity in global patent law. Berkeley Technology Law Journal 685–7: 726. http://www.btlj.org/data/articles/17_02_02.pdf. Accessed 9 May 2012

Dutfield Graham (2009) Intellectual property rights and the life science industries-past, present and future, 2nd edn. World Scientific Co. Pte. Ltd., Singapore

Grain (2001) TRIPS-plus through the back door: How bilateral treaties impose much stronger rules for IPRs on life than the WTO. http://www.grain.org/briefings/?id=6. Accessed 8 May 2012

GRAIN (2002) WIPO moves toward world patent system. http://www.grain.org/system/old/docs/wipo-patent-2002-en.pdf. Accessed 10 May 10, 2012

Habiba Saeed (2009) TRIPS: Patenting of biotechnological inventions. International Journal of Humanities 16: 64

Keim Benjamin (2007) Patent eligible subject matter in the biotechnological arts. http://works.bepress.com/benjamin_keim/1. Accessed 15 October 2011

Loughlan Patricia (1998) Patents: Breaking into the Loop. Sydney Law Review 20: 513

McManis Charles R. (2003) Patenting genetic products and processes: A TRIPS Perspective. In: F. Scott Kieff (ed.) Perspectives on Properties of the Human Genome Project. Elsevier Academic Press, London (U.K.) p. 80

Palombi Luigi (2003) Patentable Subject Matter, TRIPS and the European Biotechnology Directive, Australia and Patenting Human Genes. UNSW Law Journal Forum Genetics & Law 9: 33

Reichman, J. H. and Dreyfuss, R. C. (2007) Harmonization without consensus: critical reflections on drafting a substantive patent law treaty. Duke Law Journal 57: 86

Sekar Soundarapandian and Kandavel Dhandayuthapani (2004) The future of patent deposition of microorganisms? TRENDS in Biotechnology 22: 214

Smith Carrie P. (2000) Patenting Life: the potential and the pitfalls of using the WTO to globalize intellectual property rights. North Carolina Journal of International Law and Commercial Regulation 26: 143. Available via West Law. www.international.westlaw.com. Accessed May 2, 2012

UNCTAD-ICTSD Resource Book on TRIPS and Development (2005).Cambridge University Press, New York.

WIPO Magazine (2006) Patent Law Harmonization: What Happened? http://www.wipo.int/wipo_magazine/en/2006/03/article_0007.html. Accessed 8 May 2012

# Chapter 4
# Legal, Social and Policy Implications of Genetic Patents: Issues of Accessibility, Quality of Research and Public Health

Patenting of human gene and gene fragments has significant legal, social and policy implications as it exerts a wide range of effects on the accessibility of genetic research tools, genetic innovation, health policies, patients' rights, clinical practice and the society at large. The potential of genetic research to produce commercial results has led to the rapid commercialisation of basic genetic research through commercial agreements and patents. The commercialisation of basic genetic research has threatened the free flow and open sharing of academic knowledge. The increased commercialisation of upstream (basic) genetic research has led to patenting of gene fragments such as ESTs and SNPs, which are basically research tools. Patenting of these genetic research tools may stifle genetic innovation as a researcher has to negotiate with the patentee about the license terms before using such a research tool. Patenting of genetic testing especially in the field of diagnostics has also become a very controversial issue. Overbroad patent claims and aggressive licensing strategies stifle the innovation process. A US-based multinational healthcare company Myriad Genetics' patents on breast and ovarian cancer genes BRCA1 and BRCA2 reflect various social and policy implications involved in patenting of genetic testing.

## 4.1 Commercialisation of Genetic Research and Its Impact on Academics

Recent advances in molecular biology and genetics have changed the nature of basic genetic research. The potential of academic genetic research to produce commercial results has brought universities close to industries. Governments of different countries are supporting this trend because they believe that research carried by universities can contribute significantly in their national economy.[1] There has been enormous pressure upon publicly-funded bodies to put findings from research to commercial use. The acquisition of patents by universities is now recognised as a

---
[1] Silverstein et al. 2009.

proper indicator of academic performance. As a result, a wide range of patents is being granted to life science researchers. In the USA, more patents on DNA sequences have been granted to those carrying out research in universities than in industries.[2]

The recent trend of university-industry relationship began with the passage of the Bayh-Dole Act in the USA in 1980. The Act gave universities the right to obtain IPRs in inventions resulting from publicly funded research. It allows industries to invest and collaborate in the university research and reap exclusive benefits through licensing and commercialisation agreements.[3] In order to promote commercial development of new technologies, Congress began encouraging universities and other institutions to patent discoveries arising from federally supported research and development and to transfer their technology to the private sector.[4] In other places such as Canada and Europe, where there was no law similar to the Bayh-Dole Act, universities committed themselves to enhance their commercialisation outcomes. For instance, Canadian universities committed to triple their commercialisation outcomes by 2010 by way of the 2002 Framework on Federally Funded Research.[5] The rapid commercialisation process is increasingly diminishing the distinction between academic (non-commercial) and commercial research. Now, upstream research in the biomedical research is increasingly becoming private in myriad ways viz. supported by private funds, carried out in a private institution, or privately appropriated through patents, trade secrecy or agreements that restrict the use of materials and data.[6]

The rapid commercialisation of basic genetic research has threatened the free flow and open sharing of academic knowledge.[7] The large influx of patents, confidentiality clauses and material transfer agreements (MTAs) into academia has created serious concerns about the accessibility of research results.[8] In universities, which lack adequate resources, technology transfer skill and prestige, the MTA process can lead to unreasonable delays, acceptance of unreasonable contract provisions or abandonment of the research project altogether.[9] In the field of genetic research, protection of upstream discoveries like DNA and genetic sequences can potentially block downstream research which in turn blocks the development and access of new treatments.[10]

The commercialisation process has led to the classification between most profitable areas and not-so profitable areas. Academic researchers now focus more on most profitable areas of research. Interviews of genomic researchers conducted for a study suggests that the researchers hold the opinion that though the funding

---

[2] Nuffield Council of Bioethics, *The ethics of patenting DNA*, para. 5.32 (July 2002).
[3] Gold and Knoppers, *supra* note 1, at 133.
[4] Heller and Eisenberg 1998.
[5] Gold and Knoppers, *supra* note 1.
[6] Heller and Eisenberg, *supra* note 4.
[7] Gold and Knoppers, *supra* note 1, at 131.
[8] *Ibid.*
[9] *Id.*, at 136–137.
[10] *Id.*, at 131–132.

policies are same for all types of research, the focus of commercial research is on profit and self-sustainability rather on rapid dissemination of research results.[11]

The commercialisation process also delays the presentation and publication of research results as researchers are reluctant to share their research until the filing of its provisional patent application. They feel that if research results are published before proper filing, they can be considered 'prior art'.[12] This compels the researchers to work in a more closed environment in order to protect the value of their research at its early stages. While patenting will eventually lead to the public disclosure via publication, such disclosure is delayed throughout the research and patenting process, which can take a substantial amount of time.[13] Efforts to gain a competitive advantage can lead to longer periods of secrecy, and are thus more problematic.[14]

The purpose of commercialisation is to improve the technology transfer of cutting-edge academic research; however, the current commercialisation process may have made it more difficult for academic researchers to access and build upon scientific discoveries, and to openly disseminate their research results.[15] In a survey, majority of respondents agreed that there was a benefit to sharing data generated from basic genetic research. In this context, they viewed open access to research results as an important factor in progression of research.[16] However, in the context of the modern competitive research, all researchers were not agreed to the fact that this sharing was justified.[17] There has been a recent trend for research groups to make their results available online on a continual basis, on the condition that anyone working on the data allows the group generating it to have first publication rights.[18]

## 4.2 Importance of Patents in Genetic Research

Patents are important for innovation as they provide economic incentive to an inventor, encouraging him to invent. Once an invention is patented, the inventor alone reaps the benefit of his invention and has the right to exclude others from making using or selling his invention. Patents also encourage investment as they ensure protection of commercial benefits to investors. The economic incentive provided by patents becomes more important in fields that are new and require a lot of research and development such as biotechnology. The legal exclusivity provided by patents

---

[11] *Id.*, at 140.
[12] *Id.*, at 137.
[13] *Ibid.*
[14] *Ibid.*
[15] *Id.*, at 138.
[16] *Id.*, at 152.
[17] *Ibid.*
[18] *Ibid.*

for a certain period encourages high risk investments in new fields.[19] The quid pro quo of the patent system ensures that public must receive significant disclosure in exchange for being excluded from patenting the invention for a limited period of time.[20] In absence of patents, an inventor has no incentive to disclose an invention. In such a situation, he is left with the only option to maintain maximum secrecy, which may be a great loss to society. However, the multiple numbers of patents on a single genetic invention and the sheer breadth of patent claims relating to a gene or gene fragment have serious implications for society.

### 4.2.1 Impact of Patents with Broad Scope on Genetic Research

It is the general perception among lawyers and economists that strong and broad patent rights spur economic progress. However, concerns are being made regarding the extent of these strong patents because sometimes they create barriers to follow-up research and hinder technological advance.[21] Though it is still not very clear that what social effects the patent system creates, sceptics argue that the patent system has become an end to itself and patents could work against their original idea of innovation creation and knowledge distribution.[22]

Broad patent protection may conflict with the innovation process in emerging fields such as biotechnology, where the learning curve is particularly steep. In the biotechnology field, broad patent protection on a particular gene may limit opportunities for researchers, who want to carry on further research on that gene.[23] Therefore, DNA patents are criticized for being too broad with respect to follow-up research. Researchers are reluctant to carry on further research with a gene after knowing that it has been patented by a third party. Companies also turn down a genetic research, the prospect of which depends upon the patents of somebody else.[24]

### 4.2.2 Impact of Increasing Number of Gene Patents on Genetic Research: The Tragedy of Anticommons

In addition to patents with broad scope, it is the increasing number of patents on a single gene that creates a serious problem. Although economic theory postulates that patents facilitate the diffusion of knowledge and innovation, recent studies, however, have found that too much patenting can potentially deter innovation.[25] In

---

[19] Graham Dutfield, "DNA patenting: implications for public health research" 84 *Bulletin of the World Health Organisation* 388 (2006).

[20] Zekos 2005.

[21] Thumm 2005.

[22] *Id.* at 1411.

[23] Dutfield, *supra* note 19 at 389.

[24] Thumm, *supra* note 21, at 1411.

[25] *Ibid.*

the USA, Heller and Eisenberg described this situation as 'the tragedy of anticommons'. They referred to the metaphor 'tragedy of the commons' used by Garrett Hardin to explain the overuse of common resources. People often overuse resources they own in common due to lack of incentive to conserve them. This provides powerful justification for privatising the common property. Heller and Eisenberg, however, criticized the metaphor by saying that while it takes into account the overuse of a scarce resources by people, it overlooks the other possibility of underuse of the same when governments give too many people rights to exclude others.[26] They explain anticommons property as the mirror image of commons property:

> A resource is prone to overuse in a tragedy of the commons when too many owners each have a privilege to use a given resource and no one has a right to exclude another. By contrast, a resource is prone to underuse in a "tragedy of the anticommons" when multiple owners each have a right to exclude others from a scarce resource and no one has an effective privilege of use.[27]

In the biomedical field, multiple patents on genes and gene fragments, which can be used as a tool for further research, discourage genetic innovation. Access to these tools demands negotiation with all the patent holders, which can raise the transaction costs. This may result in underuse of resources.

### 4.2.3 Patent Thickets

Multiple patents on genes and gene fragments may lead to other problem i.e. patent thickets. While anticommons problem requires the aggregation of multiple inputs to create a single product, patent thickets occur when multiple overlapping patents cover the same technology and choke an industry. Patent thickets create an environment where patent holders can prevent each other from fully utilising the corresponding rights. Here, each holder's right overlaps with, and infringe upon, a right held by another.[28] Therefore, patent thickets create a dense web of overlapping patents, where every patent holder has to find its way through the web in order to exercise his rights effectively. This again requires negotiation of multiple licenses, which may increase the transaction costs and retard the pace of innovation.[29]

### 4.2.4 Royalty Stacking

A considerably large number of patents on genes and gene fragments may lead to royalty stacking problem. The concept of royalty stacking stems from the risk that

---

[26] Heller and Eisenberg, *supra* note 4.
[27] *Ibid.*
[28] Safrin 2004.
[29] Thumm, *supra* note 21, at 1411.

multiple patents may affect a single product.[30] Such risk is relatively high in the biotech field which is flooded with patent filing. Royalty stacking arises when, in order to take a product to market, the developer of the product takes licences from all of the owners of the patents which affect the final product. When the royalty payments are added together, the licensee may find itself with a non-profitable product.[31] The royalty stacking problem would have an adverse impact on further research and development. It would also discourage or misdirect research, slowing down the development of socially beneficial products and processes.[32]

## 4.3 Patenting of Genetic Research Tools and its Impact on Research and Innovation

A disproportionally large number of patents are being granted with respect to genes as compared to the number of commercial products based upon them. The underlying reason behind this is that among these patents, an enormous quantity of patents is being granted on genes and gene fragments that are basically research tools.[33] Here, the concern is that the future commercial products such as therapeutic proteins and diagnostic tests require the use of multiple research tools, many of which are patented. In this situation, patent holder of genetic research tools often charge exorbitant licensing fee and users of genetic research tools incur prohibitive transaction costs.[34]

Although, there is no precise definition of the term research tools, however, it can be defined as 'resources used by scientists that have no immediate therapeutic or diagnostic value, but have great value in conducting scientific work'.[35] It can also be defined as a tool (including both composition of matter and methods) that is used in a laboratory, not for the purpose of improving itself but to achieve a certain research goal.[36] In other words, it is to be used principally as a means of developing a commercial product, such as a medicine or vaccine, rather than constituting a product in itself.[37] Genetic research tools, which are being patented, may consist of entire genes, parts of a gene or a few base pairs. The encoded products of the gene

---

[30] Vicky Clark, "Pitfalls in drafting royalty provisions in patent licences" *Bio-Science Law Review*, available at http://pharmalicensing.com/public/articles/view/1087832097_40d70021d738c (last visited on Dec 20, 2011).

[31] *Ibid.*

[32] Thumm, *supra* note 21.

[33] Dutfield, *supra* note 19, at 389.

[34] *Ibid.*

[35] Gold et al. 2005.

[36] Sumikura 2009

[37] Nuffield Council, *supra* note 2, para. 5.30 (July 2002).

are not always fully elucidated as they may have the potential to yield commercial products in future when the function is better understood.[38]

Two important research tools are expressed sequence tags (ESTs) and single nucleotide polymers (SNPs). In the development process of ESTs, the coding parts of genes could be rapidly sequenced; this led to the extensive application of this method as a means of locating entire genome. Now, when the sequencing of the human genome has been completed and many genes of unknown function are available for study, access to these data is accelerating our understanding of disease. The second important research tools, SNPs, are used in research to help locate genes associated with disease or identify genetic variation which may predispose to disease.[39] The biotechnology and pharmaceutical companies are the main users of genetic research tools in the discovery and development of new medicines. With the understanding of the role of the products of particular genes and their mutations in cellular pathways, it can be possible to modify their effects through the action of medicines.[40]

### 4.3.1  Patentability of Genetic Research Tools

There has been a consensus among the legal world that genetic inventions, including genetic research tools are patentable. The joint statement by the three major patent offices: EPO, USPTO and Japan Patent Office, took the following position:

> Purified natural products are not regarded under any of the three [European, US and Japanese] laws as products of nature or discoveries because they do not in fact exist in nature in an isolated form. Rather, they are regarded for patent purposes as biologically active substances or chemical compounds and eligible for patenting on the same basis as other chemical compounds.[41]

Germany's purpose-bound protection limits claims over human genetic inventions to particular identified uses. So a second use of a patented human DNA sequence which is not patented is not an infringement. This ensures the use of a patented human genetic sequence for second use.[42]

### 4.3.2  Implications of Patents Relating to Genetic Research Tools for Society

Gene patents may have an adverse impact on drug discovery because drug industries conduct screening of the compounds. New drugs are developed on the basis of

---

[38] *Ibid.*
[39] *Id.*, at para 5.31.
[40] *Ibid.*
[41] Gold, Joly & Caulfield, supra note 35 at 2..
[42] *Ibid.*

such screening of compounds which involves use of various research tools such as cell lines, analysers, research reagents, research kits etc. Here, patenting of genetic research tools can influence the development of drugs and improvement of medicines.[43]

Since it is an accepted principle of the patent system that the monopoly awarded to an inventor should reflect the contribution that he has made, therefore, the Nuffield Council recommends that the claims in a patent that asserts rights over a DNA sequence which has a use in drug screening should not 'reach through' to the product. By 'reach through', we mean the capacity the owner of a patent has to claim rights over further unrelated uses identified by researchers at a later stage.[44]

### 4.3.3 Patenting of ESTs and Reach Through Claims

In a case, when rights in relation to a partial DNA sequence or EST are asserted in a patent, there is a possibility that the patent will also extend to the full DNA sequence, despite the fact that the full sequence may be isolated by someone else without using the EST in question. This possibility of 'reach through' by ESTs to whole genes has been one of the principle concerns of those who have objected to the idea that ESTs may be patentable. There is a broad consensus world over as to the fact that patent protection of partial DNA sequences such as ESTs should not be granted in broad terms. Nuffield Council recommends that 'when rights are asserted in terms intended to cover all sequences that contain the EST that is the subject of the original patent, no patent should be granted.'[45]

### 4.3.4 Impact of Patenting of Genetic Research Tools on Innovation

The grant of increasing number of patents would increase the cost of research as more licenses would be required to conduct research. In such a situation, conducting research may become more difficult because researchers would be required first to negotiate the use of patented genes and sequences. Furthermore, a patent owner may withhold a licence to gain maximum financial benefits, or licence it exclusively to one or a limited number of licensees. If a company needs to acquire the rights to several DNA sequences to develop a therapeutic protein or diagnostic test, the company may decide not to develop such protein or test because of the number of royalty payments that would be required (this is sometimes referred to as royalty-stacking).[46] However, recent studies reveal no sufficient evidence, which indicates

---

[43] Sumikura, *supra* note 36 at 74–75.
[44] Nuffield Council, *supra* note 2, para. 5.37.
[45] Nuffield Council, *supra* note 2, para. 5.38.
[46] *Id.* para. 5.39.

that the patenting over genetic research tools have a potentially deleterious effect on upstream research.[47]

Nuffield Council took the view that 'the exercise of a monopoly over what are now essentially discoveries of genetic information accessible by routine methods is, in principle, highly undesirable.'[48] It opined that 'in general, the granting of patents which assert rights over DNA sequences as research tools should be discouraged and the best way to discourage the award of such patents is by a stringent application of the criteria for patenting, particularly that of utility.'[49] It has welcomed Utility Guidelines for DNA sequences introduced by the United States Patent and Trademark Office (USPTO), which have, in effect, been endorsed by the European Patent Office (EPO).[50]

The council further recommended that 'the United States Patent and Trademark Office (USPTO), the European Patent Office (EPO) and the Japan Patent Office (JPO) should monitor the impact of the Guidelines on the examination of patents to ensure that the criterion for utility is rigorously applied so that the grant of a patent more properly reflects the inventor's contribution. If this proves not to be the case, the Guidelines should be reviewed and strengthened to achieve this purpose.'[51] As discussed earlier, multiple patents on genetic research tools may lead to another problem of patent thickets.

## 4.4 Common Practice Regarding Using Patented Research Tools in Public Sector and in Private Sector

*Patent Infringement and Anxiety Among Researchers and Academicians to Use Patented Research Tools* As a common practice within the research community, researchers freely use research tools for academic purposes, provided patents for such tools are owned by an academic institution or individual researchers belonging to such an institution. Furthermore, even in cases, where a certain research tool is patented by a private sector company, the risk that researchers may face litigation as a result of using the tool in academic research has actually been extremely low. This is because 'private sector companies see little merit in exercising their rights against the use of their patented research tools in the academic research.'[52] However, the interviews conducted by the Koichi Sumikura from 2004 to 2005, revealed the anxiety among researchers by using research tools in academic research has risen

---

[47] *Id.* para. 5.40.
[48] *Ibid.*
[49] *Id.* para. 5.41.
[50] *Ibid.*
[51] *Ibid.*
[52] Sumikura, *supra* note 36 at 77–78.

to unprecedented level.[53] Patents on research tools are rarely enforced and firms holding patents on research tools rarely sue public research bodies for infringement. Research projects are very rarely halted due to patent issues because in most cases, working solutions such as research exemptions, reach through claims etc. are found.[54]

A survey conducted by John P. Walsh and others reveals that university and industrial researchers have adopted 'working solutions' that allow their research to proceed. These include licensing, inventing around patents, going offshore, the development and use of public databases and research tools, court challenges, and simply using the technology without a license (i.e. infringement).[55] Licensing has become a routine thing in the drug industry, where, problem of access to patented research tools or upstream discoveries can often be settled contractually.[56]

*Patent Infringement by Researchers by Invoking Research Exception Justification and Tolerance by the Companies Having Patents* The survey also reveals that infringement of research tool patents, particularly by university researchers, is common.[57] Researchers, however, justify such infringement by invoking a 'research exemption'. At the other end, industrial IP holders tolerate the infringement of their patents by academic researchers (with the exception of patents on diagnostic tests used in clinical research), partly because it can increase the value of the patented technology.[58] Further, industrial patent holders consider that the small prospective gains from a lawsuit do not worth the legal fees with the attendant risk of the patent being narrowed or invalidated and bad publicity from suing a university. They are also reluctant to 'upset the norms of open access in this community of academic and industrial researchers for fear of losing the goodwill of one's peers and the associated access to materials and information'. Despite these concerns, the firms interviewed for the survey, were willing to defend against competitors' infringement of their core patents, especially those on potential therapeutics.[59]

## 4.5 Viable Options

### 4.5.1 *Exclusive Licensing Practices May Retard Innovation*

Many organisations including universities and biotechnology companies, which have been granted patents in relation to genetic research tools, are often not well

---

[53] *Id.* at 78.
[54] Thumm, *supra* note 21 at 1442.
[55] Walsh et al. 2003.
[56] *Ibid.*
[57] *Ibid.*
[58] *Ibid.*
[59] *Ibid.*

placed to undertake extensive product development and distribution. Therefore, they often seek to realise the value of their patents through licensing. In this situation, there is always a risk that an important patent may be licensed exclusively. Such exclusivity may not be in the public interest as it may discourage others from working in an area which would profit from a variety of approaches or solutions.[60]

### 4.5.2 Non-exclusive Licensing over Genetic Research Tools Should be Encouraged

Nuffield Council recommend that 'those public institutions which already have been awarded patents that assert rights over DNA sequences as research tools be strongly encouraged not to licence them exclusively to one or a limited number of licencees, even when, by not doing so, they may suffer some loss of revenue in the short term.'[61] It also recommends that, 'wherever possible, the private sector should consider nonexclusive licensing for those DNA sequences which are used in research.'[62]

### 4.5.3 Research Exemptions and Their Scope

Research exception[63] is seen as the most straightforward and viable option to meet the challenges of accessibility. Emphasising on the need for research exception in the context of human genetic research, Human Genome Organisation (HUGO) states that 'freedom to undertake research in genomics is fundamental to the acquisition of further knowledge and the development of applications to benefit human health'[64] and that 'researchers who wish to use patented technologies or products to further understanding in their work should not be unduly constrained by issues relating to licensing.'[65] Most countries already possess either a statutory or a com-

---

[60] Nuffield Council, *supra* note 2, para. 5.42.
[61] *Ibid.*
[62] *Ibid.*
[63] Nuffield Council, *supra* note 2, para. 5.43; (Researchers sometimes seek to utilise patented DNA sequence in research without any obvious prospect of commercial development arising from that use. This situation occurs in the context of most academic research, as well as some research in industry because research may be undertaken on inventions which have been patented, including DNA sequences. This is generally termed as the 'research exemption.' Patent systems of most of the countries have some form of exemption 'to enable research to be carried out on a patented invention provided it is not intended to produce commercial benefit, so as to ensure that innovative research is not stifled.').
[64] Gold, Joly and Caulfield *supra* note 35 at 3; quoting HUGO Intellectual Property Committee, *Statement on the Scope of Gene Patents Research Exemption, and Licensing of Patented Gene Sequences for Diagnostics* (London: The Human Genome Organization, 2003).
[65] *Ibid.*

mon law research exception but their scopes differ markedly from jurisdiction to jurisdiction. In addition to this, 'the variety and imprecision of the language used to define the exception also makes it difficult for courts of justice to ensure judicial stability.'[66]

**Research Exception in the USA** In the USA, Congress enacted the Hatch-Waxman Act in 1984, which incorporated a limited industry-specific experimental use exception for testing drugs and medical devices for purposes reasonably related to regulatory data gathering.[67] Courts interpreted the term 'reasonably related' widely to include food additives, colour additives, new drugs, antibiotic drugs and human biological products.[68] In a recent case, *Integra v. Merk*,[69] the CAFC seems to have limited the scope of this statutory exception to experiments carried out for the purposes of facilitating expedited marketing approvals for generic drugs only. The court reasoned that extending such an exception to new drug development would be contrary to the purpose and language of Hatch-Waxman Act.[70] The Hatch-Waxman Act has, therefore, established the right of generic pharmaceutical manufacturers to make use of the patented product prior to the patent expiry date and without the approval of the patent holder, for the purpose of engaging in preparatory acts with a view to obtain marketing approval from drug regulatory authorities upon expiry of the patent.[71] Such use facilitates the entry of generic competition as soon as possible after the date of patent expiration; otherwise a generic competitor would only be able to start its bioequivalence and other testing only after patent expiry, which would result in the de facto extension of patent protection.

The common law experimental use exception was introduced in US patent law for the first time by Justice Story in *Whittemore v. Cutter*[72] in early nineteenth century.[73] Justice Story maintained that 'it could never have been the intention of the legislature to punish a man, who constructed such a machine merely for philosophical experiments, or for the purpose of ascertaining the sufficiency of the machine to produce its described effects.'[74] In the same year, *Sawin v. Guild*[75] widened

---

[66] Gold, Joly and Caulfield, *supra* note 35 at 1–2.

[67] "Patenting and the Research Exemption," available at http://www.transknowlia.org/content/patenting-and-research-exemption (last visited on August 12, 2012); See 35 U.S.C. Sec. 271 (e).

[68] *Ibid.*; See *Eli Lilly and Co. v. Medtronic, Inc.* 496 U.S. 661 [1990]; *Intermedics, Inc. v. Ventritex Inc.*, 775 F. Supp. 1269 (N.D. Cal. 1991).

[69] *Integra Life Sciences I, Ltd v. Merck KGaA*, 2003 U.S. App. LEXIS 11335 [Fed. Cir. 2003].

[70] *Supra* note 5.

[71] See the U.S. Drug Price Competition and Patent Term Restoration Act (the Hatch-Waxman Act), 1984.

[72] *Whittemore v. Cutter*, 29 F. Cas. 1120 (C.C.D. Mass. 1813) (No. 17, 600).

[73] Evans Misati and Kiyoshi Adachi, "The Research and Experimentation Exceptions in Patent Law: Jurisdictional Variations and WIPO Development Agenda" *UNCTAD-ICTSD Project on IPR and Sustainable Development*, available at http://unctad.org/en/docs/iprs_in20102_en.pdf (last visited on August 12, 2012.)

[74] *Whittemore, supra* note 10.

[75] *Sawin v. Guild*, 21 F. Cas. 554, 555 (C.C.D. Mass. 1813) (No. 12,391).

## 4.5 Viable Options

experimental use beyond machines and also introduced the concept of non-commercial use when the court excluded from infringement the exploitation of a patented invention unless it constituted 'making with an intent to use for profit, and not for the mere purpose of philosophical experiment, or to ascertain the verity and exactness of the specification.'[76] This exception, however, was held to be 'truly narrow' in *Roche v. Bolar*[77] and the slightest commercial purpose or intention for carrying out an experiment has been held to be patent infringement.[78] Such narrow interpretation of the experimental use exception has also taken its toll on non-state university research.[79]

State universities are apparently immune from infringement liability of federal patent laws as held by the US Supreme Court in *Florida v. College Savings Bank*.[80] The decision was made on the grounds of the Eleventh Amendment immunity from patent infringement liability for states, instrumentalities of states and state employees acting in their official capacity.[81] Duffy comments that 'this judgement may lead patentees to bring patent infringement suits before the state courts under the guise of unfair competition actions and secondly, the effect of state universities being granted immunity could lead to the private sector industry utilising state universities for carrying out research, which would otherwise require a license from patentee.'[82]

This position, however, was changed in the case *Madey v. Duke University*.[83] In this case, Duke University employed Madey as a laboratory director. He owned two patents, before his appointment at the University. After the termination of his services, Madey sued Duke University for infringement of his patents. Duke University relied on, inter alia, the experimental use exception defence. The CAFC refused to apply the experimental use exception to exempt university research activities from infringing a patent by maintaining that these research activities 'unmistakably further the institution's legitimate business objectives, including educating and enlightening students and faculty participating in these projects'. The court also disregarded the non-profit status of Duke University. The rationale behind the Duke University judgement could be found in the 1980 enactment of the Bayh-Dole Act which aims at promoting collaboration between commercial concerns and non-profit organisations, including universities.[84]

---

[76] Misati and Adachi, *supra* note 11.

[77] *Roche Products, Inc v. Bolar Pharmaceuticals* Co., Inc., 733 F.2d 858 [Fed. Cir. 1984].

[78] *Supra* note 5; *See Pfizer Inc. v. International Rectifier Corp.*, 217 USPQ 157 (C.D. Cal. 1982); *Embrex, Inc. v Service Engineering Corp.*, 216 F.3d 1343 (Fed. Cir. 2000).

[79] *Ibid.*

[80] *Florida Prepaid Postsecondary Education Expense Board v College Savings Bank*, 527 U.S. 627 (1999).

[81] *Supra* note 5.

[82] *Ibid.*; *See* Duffy 2002.

[83] *Madey v Duke University* 307 F.3d 1351 (Fed. Cir. 2002).

[84] *Supra* note 5. (It encourages universities to patent 'subject inventions' (means any invention of the contractor conceived or first actually reduced to practice in the performance of work under a government contract, as defined in Federal Acquisition Regulations-FAR, Sec. 27.301), made with federal government funds and contemplates them to grant licenses of patents to the private sector.).

According to Mueller, 'legal, technological and economic factors have removed the strict barriers between academic research and commercial research, thus pure "philosophical" experimentation is no longer possible, calling for a broader interpretation of the common law experimental use exception to ensure that university research does not get stifle out.'[85] However, the AIPPI (International Association for the Protection of Intellectual Property Rights) report of the American group on 'Patentability Requirements and Scope of Protection of Expressed Sequence Tags (ESTs), Single Nucleotide Polymorphisms (SNPs) and Entire Genomes'[86] has suggested no change to the experimental use exception for biotechnology law.[87] The current position in the USA is that experimental use exception to patent rights in the USA is 'truly narrow' and non-state academic institutions are not exempted from patent infringement liability as such.[88] Although the USA, among others, had a long tradition of cases emphasising that only non-commercial research is exempt from patent infringement liability; however, in practice, it is often difficult to delineate between commercial and non-commercial research and experimentation.

The USA, among others, had a long tradition of cases emphasising that only non-commercial research is exempt from patent infringement liability. While not referring to non-commercial research as such, Section 47(3) of the Indian Patent Act stipulates that acts that constitute 'merely of experiment or research' (emphasis added) are exempt from patent infringement. In practice, it is often difficult to delineate between commercial and non-commercial research and experimentation. A number of factors have blurred the line between research that advances legitimate business (commercial) and research that is purely academic or non-commercial. One factor is the way in which research is conducted, since applied commercial research relies on basic research done in universities and other research institutions.[89] Other factors include legal developments such as the Bayh-Dole Act[90] and similar acts in other countries that encourage academia to apply for patents on their research to enable the commercialisation of innovation.[91] This blurring has perhaps led to a narrowing of the research exception in some of these countries. In the USA, the 2002 *Madey v. Duke* ruling maintained that experimental research, using a patented product without the consent of the patent holder, constitutes patent infringement where used to further 'the infringer's legitimate business' interests. This ruling is widely seen as having curtailed the defence that universities, whose charters committed their institutions to pursue a non-profit objective, enjoyed a wide research exception defence against claims of patent infringement.[92]

---

[85] *Ibid.*; *See* Mueller 2001.

[86] Available at www.aippi.org.

[87] *Supra* note 5.

[88] *Ibid.*

[89] Misati and Adachi, *supra* note 11.

[90] 35 U.S.C. Sec. 200–212.

[91] See, for example, in India, The Protection and Utilization of Publicly Funded Intellectual Property Bill (2008).

[92] Madey v. Duke University, 307 F.3d 1351 (Fed.Cir. 2002), cert. denied 539 U.S. 958, 123 S.Ct. 2639, 156 L.Ed.2d 656 (2003).

**Research Exception in the European Union** The extent to which experimental use of patented inventions is permitted in Europe varies among member states of the European Union and depends upon national patent laws. Art. 64(1) of the European Patent Convention (EPC) provides that the rights conferred by a European patent in all designated countries to which the European patent extends shall be the same as those conferred by a national patent granted in that state.[93] Article 64 (3) of the EPC provides that any infringement of a European patent shall be dealt with by national law.[94] Thus, no provision regarding defences to infringement is found in the EPC.

The European Union Directive 2004/27/EC exempts acts done for regulatory approval purposes.[95] Many countries have moved to adopting a separate exception in the context of pharmaceutical clinical trials. This exception is known as the regulatory review exception and is also called the 'early working' or 'Bolar' exception, referring to a case involving a party of the same name.[96] Art. 9(b) of the draft Community Patent Regulation of 2004 states, 'Community Patent shall not extend to acts done for experimental purposes relating to the subject matter of the patented invention.'[97] Unlike the USA, all the member states of the European Union, except Austria, have introduced a general non-industry specific experimental use exception in their patent statutes.[98]

According to Benyamini,[99] the legislative intent of the experimental use exception, as debated during the drafting of Art. 27(b) of the Community Patent Convention[100] shows distinction between the private use exception and experimental use exception.[101] Furthermore, the drafters intended that the experimental use exception would exclude experiments carried out for industrial and commercial purposes. The drafters intended to maintain a distinction between academic research and industrial research for the purposes of the experimental use exception to patent rights.[102] However, according to Cornish, the form of words ultimately adopted in the CPC and state legislations have failed to distinguish academic research from research in the industry and the courts of member states have declined to follow such a distinction.[103]

Initially, European countries restricted their research exemption to non-commercial activity, i.e. typically, to work in universities and public institutions without

---

[93] Art. 64 (1) of the European Patent Convention (EPC).

[94] *Id.* Article 64(3).

[95] Misati and Adachi, *supra* note 11; *See* the European Union Directive 2004/27/EC.

[96] *Ibid.*; *See Roche Products Inc. v. Bolar Pharmaceutical Co.* (733 F. 2d. 858, Fed. Cir., cert. denied 469 US 856, 1984).

[97] Art. 9(b) of the draft Community Patent Regulation of 2004.

[98] *Supra* note 5.

[99] Reported in Cornish 1998.

[100] Art. 27(b) enunciates: "[t]he rights conferred by a Community patent shall not extend to acts done for experimental purposes relating to the subject-matter of the patented invention."

[101] *Supra* note 5.

[102] *Ibid.*

[103] *Ibid.*; *See* Cornish 1998.

industrial backing. Under the present law, there are separate provisions which on the one hand exempt use which is private and non-commercial, and, on the other hand, experimental use. In consequence, courts across Europe have shown increased willingness to treat experimental research as exempt from patent liability even though it has a commercial purpose.[104]

In the context of biotechnology patents, ambiguity persists regarding how far clinical tests can be regarded as experimental because treatment and the continuing search for further genetic knowledge often enough go hand in hand. It would be good that these tests be treated as exempted tests only when the latter objective is a dominant motive for the tests but the law remains rather uncertain.[105]

Furthermore, there has been a paradigm shift in the roles governments play. In the present time, governments have decreased their roles in 'non-core sectors', which incidentally also include education, thereby encouraging the private sector to invest into the education sector. As a result, private and non-state universities have emerged as a new phenomenon in the 'business' of providing education. The present universities are actively involved in contract research for the private sector. Most universities in Europe have established technology transfer organisations internally or engage companies to hold, manage, license and enforce their intellectual properties. In the USA, university patentees are extremely zealous in guarding and enforcing their patent rights, which too can perhaps justify the *Madey v. Duke University* judgement.[106]

Highlighting the difficulty in defining the scope of research exception, Evans Misati and Kiyoshi Adachi comment:

> ...At the policy level, a typical means of attempting to distinguish between research activities that fit into the exception and those that do not is to decide whether the research is essentially of a commercial or non-commercial nature. In practice, however, it is becoming increasingly difficult to make such a distinction. In the US, for instance, the trend has been for courts to narrow the exception, since much scientific research could be said to have some commercial aspect, as in the case of research conducted by universities (i.e., the Madey case).[107]

**Statutory Exception in Developing Countries** The statutory exception of a number of countries did not link experimental use to non-commercial purposes. For the most part, these countries treat the private/non-commercial condition as an independent exception to patent infringement.[108] Certain developing countries such as India, Brazil, Argentina and Uruguay had broad statutory research exceptions that did not require the experimental use to be 'related' to the subject matter of the patented invention. However, due to lack of legal precedents on the topic of the

---

[104] *Ibid.*; For the UK, see *Monsanto v. Stauffer* [1985] RPC 515. For recent confirmation of the new approach in France, *Wellcome Foundation v. Parexel International & Flamel*, Tribunal de Grande Instance de Paris, 20th February 2001: *Intellectual Property News* Issue 17, July 2001.
[105] *Ibid.*
[106] *Ibid.*
[107] *Ibid.*
[108] *Ibid.*

research exception in those countries it was difficult to foresee how the courts will interpret these broad exceptions.[109]

The Bolar exception[110] has been maintained by many countries in their patent legislation. India, which is the largest supplier of generic medicines, also maintains a Bolar exception in Section 107 of its Patent Act. In this regard, it should be noted that the European Community, which previously opposed a regulatory review exception, has adopted a version of its own.[111]

The uncertainty as to the language and scope and the changing nature of genetic research creates some obstacles in the accessibility of research tools. Nevertheless, broad research exception is the most straight forward option before the researchers and innovators. There is a dire need to define these research exceptions in the light of current technological developments and changing scenario of research.

## 4.6 Patenting of Genetic Tests for Diagnostic Purposes

The identification of a faulty gene responsible for a disease and DNA sequences implicated in a disease provides the basis for a diagnostic test.[112] In recent years, many genetic mutations responsible for causing disease have been identified and used as a basis for clinical diagnosis. Mutations in the gene causing cystic fibrosis and haemochromatosis have been the subject of patents relating to diagnostic tests.[113] These diagnostic tests intend to alert patients and their doctors to a pre-disposition to major diseases.[114] Diseases, which involve only single gene mutations, are infrequent as majority of common diseases and disorders appear to be more complex. The majority of common diseases depend upon multiple factors including a number of genes and environmental factors. This makes the identification process more complex and cumbersome. Further, since these diseases depend upon multiple factors, it is difficult to provide full and reliable prediction about the disease.[115] Since gene-based diagnostic testing is a relatively new area, research and development in this area requires huge investment backed by strong legal protection. Therefore, despite the fact that these tests lack full and reliable prediction about the disease, investments are being made to develop a new generation of diagnostic tests.[116] The

---

[109] *Id.* at 5.

[110] An exception to patent rights allowing a third party to undertake, without the authorisation of the patentee, acts in respect of a patented product necessary for the purpose of obtaining regulatory approval for a product., available at http://www.iprcommission.org/papers/text/final_report/appendixhtmfinal.htm.

[111] *Ibid.*; *See* Directive 2001/83/EC, as amended by Directive 2004/27/EC.

[112] Nuffield Council, *supra* note 2, para. 5.4.

[113] *Id.* 5.5.

[114] *Id.* para 5.6.

[115] Nuffield Council, *supra* note 2, para. 5.5.

[116] *Id.* At 5.6

investments in this field heavily rely on the protection of the patent system because it gives control over the use of the DNA sequences for more complex and more important uses– such as identifying biochemical pathways in disease and drug targets within those pathways.[117]

Although patent is necessary for the development of this field, however, there is another face of the patenting of genetic research tools, which stifles the innovation process. Serious concerns are being made over the current patent practices relating to genes, which are important in diagnosis, management and risk assessment of human disease. The enforcement of patents can be restricted by monopolistic licensing that limits a given genetic test to a single laboratory, by royalty-based licensing agreements with exorbitant up-front fees and per-test fees, and by licensing agreements that seek proportions of reimbursement from testing services.[118] It has a far reaching implication beyond patient care as it affects the training of the next generation of medical and laboratory geneticists, physicians and scientists in the area covered by the patent or license.[119]

Myriad case presents an example for anti-competitive practices with genetic testing patents. The patent held by Myriad Genetics protects the isolated gene, as such (a chemical molecule), and the corresponding protein, and it includes the imaginable future therapeutic uses of the BRCA1 gene. The patent is broad and dominant since it implies that any other patent application filed for a different use of BRCA1 is dependent on the patent held by Myriad Genetics.[120] Myriad Genetics has obtained the exclusive rights to diagnostic tests for BRCA1 and BRCA2 in many countries. The company follows an aggressive licensing strategy against any laboratory using the tests. The approach of the company has been criticized worldwide.[121] Since Myriad' patents on BRCA1 and BRCA 2 have been discussed most worldwide, therefore, it becomes pertinent to discuss these in detail.

### 4.6.1 Myriad's Patents on BRCA1 and BRCA2 Genes: A Case Study

Among all breast cancers, hereditary breast cancer accounts for approximately 5–10%. Mutations in BRCA1 and BRCA2 are thought to be responsible for the overwhelming majority of inherited cases.[122] The BRCA1 gene is a large nucleoprotein that lies on chromosome 17. More than 700 mutations of BRCA1 have been discovered to pre-dispose to the development of both breast and ovarian cancer.

---

[117] *Ibid.*
[118] Thumm, *supra* note 21 at 1414.
[119] *Id.* at 1415.
[120] *Id.* at 1414.
[121] *Ibid.*
[122] Paradise 2004.

4.6 Patenting of Genetic Tests for Diagnostic Purposes 155

BRCA2 is a much larger gene lying on chromosome 13, with over 800 reported mutations, many often associated with early onset breast cancer.[123]

Myriad Genetics was founded by Mark Skolnick and Nobel Laureate Walter Gilbert in 1993 with the objective to develop diagnostics and to establish a world-wide business based on the discovery and commercialisation of genes linked to major disorders such as cancer and heart disease. While focussing primarily on breast and ovarian cancer, Myriad attempts to find genes for prostate, lung and colon cancer, obesity and hypertension. The company seeks to capitalise its discoveries by providing testing and genetic information services and develop human therapeutic products independently and in conjunction with commercial patents.[124]

Dr. Mark Skolnick and Myriad Genetics had become successful in isolating and locating the genes BRCA1 and BRCA2 by utilising the genealogical records of the Utah Mormons, and the processing power of super-computers. Myriad Genetics filed patent application on the discovery, including a 'composition of matter' patent on the gene itself and a 'method-of-use' patent for the application of BRCA1 and BRCA2 in the diagnostic and therapeutic arenas. Skolnick extends his support to these patents with a belief that these patents are vital in order to encourage private investment and entrepreneurship, promoting the growth of the high quality inexpensive test.[125]

Myriad has also filed a number of other patents in the USA.[126] It has also obtained a number of foreign patents covering BRCA1 and BRCA2 breast and ovarian cancer genes and their use in the development of therapeutic and predictive medicine products. The company has obtained patents in Europe, Canada, Australia, New Zealand and Japan.[127] Myriad Genetics has promised that its commercial genetic predisposition testing is a boon for patient care and the delivery of health care.[128]

### 4.6.2 Concerns Regarding Myriad's Patents on BRCA1 and BRCA2 Genes: Reactions Against Commercial Testing in the USA, European Union and Canada

**The USA** There has been significant opposition against the commercial BRCA testing and against the Myriad's monopoly in the concerned field. While criticizing the Myriad's public education program, various support groups such as Breast Cancer Action have been charging that it has far more to do with increasing anxiety and convincing women and their physicians of the need for testing, than actually informing people of the facts about breast cancer. In the light of these growing

---

[123] *Ibid.*
[124] Rimmer 2008.
[125] *Id.* at 187–188.
[126] *Id.* at 188.
[127] *Ibid.*
[128] *Ibid.*

concerns regarding BRCA genes, the American College of Medical Geneticists has called for a ban on human gene patenting, arguing that it leads to monopolistic licensing and exorbitant fees.[129]

In addition to this, 'in March 2002 hearings were held by the U.S. Federal Trade Commission[130] and a sub-committee of the House of Representatives' judiciary committee to address issues with regards to biotechnology and patent policies; in the same month, a bill was introduced (by Democratic representative Lynn Rivers) that would require the Director of the Office of Science and Technology Policy to study the impact of government policies on the innovation process for genomic technologies, and would create an exemption from patent infringement for researchers and clinicians who use genetic-based diagnostic tests for non-commercial purposes.'[131] In addition to this, in 2007, Congress introduced a bill, the Genome Research and Accessibility Act (GRAA)[132], that would have radically altered US patent law, banning patents on the human genome and naturally occurring genes as well as synthetic DNA or RNA molecules.[133] GRAA was motivated by the concern that human genes patents threaten to impede biomedical research. However, GRAA did not garner much support and after its introduction and referral to subcommittees in early 2007 the Act seems to have been abandoned.[134] The Secretary's Advisory Committee on Genetics, Health and Society (SACGHS) for the Department of Health and Human Services (HHS) has recently produced a report on gene patents, in which, it proposes to exempt healthcare practitioners and researchers from infringement liability for gene patents.[135]

There has been an organised opposition against the ill effects of gene patents on society. This is reflected in the *Association of Molecular Pathology v. USPTO*[136] case, where a lawsuit was filed on behalf of researchers, cancer survivors, breast cancer and women's health groups, and scientific associations representing geneticists, pathologists and laboratory professionals against the Myriad Genetics Inc in the New York District Court. In this case, regarding the method claims District Court Judge Sweet held that comparisons of DNA sequences involved in diagnostic gene patents are abstract mental processes, therefore also unpatentable. Echoing

---

[129] Williams 2002; See American College of Medical Genetics, *Position Statement on Gene Patents and Accessibility of Gene Testing* (ACMG, 2 August 1999), available at http://www.faseb.org/genetics/acmg/pol-34htm.

[130] Federal Trade Commission, *Competition and Intellectual Property Law and Policy in the Knowledge Based Economy*, available at http://www.ftc.gov/opp/intellect/index.htm.

[131] *Id.* at 137–138.; *See* Genomic Research and Accessibility Act of 2002 H.R. 3966.

[132] H.R. 977, 110th Cong. (2007).

[133] Caity Ross (edited by Abby Lauer), "Digest Comment: The Future of Gene Patenting after Association for Molecular Pathology v. U.S. Patent and Trade Mark Office, *Journal of Law and Technology Digest*, posted on May 14, 2010, available on http://jolt.law.harvard.edu/digest/patent/digest-comment-the-future-of-gene-patenting-after-association-for-molecular-pathology-v-u-s-patent-and-trademark-office (last visited on May 10, 2012)

[134] *Ibid.*

[135] *Ibid.*

[136] Civil Action No. 09-4515 RWS (S.D.N.Y. 2009).

## 4.6 Patenting of Genetic Tests for Diagnostic Purposes

a similar approach, the CAFC explicitly invalidates the comparing and analysing method claims which have been notoriously exercised by Myriad.[137] The Leahy-Smith America Invents Act 2011 addresses the problem of aggressive patent licensing relating to genetic testing, which prevents patients to seek second opinion about their disease. Under Sec. 23 of the Act, study on genetic testing requires the director to conduct the study on effective ways to provide independent, confirming genetic diagnostic test activity where gene patents and exclusive licensing for primary genetic diagnostic tests exist. A report to Judiciary Committees is required within 9 months with appropriate findings and recommendations.[138]

Like the USA, significant opposition to commercial genetic testing and, particularly, the patenting of BRCA genes has also developed in Europe and Canada as Myriad has obtained patents and begun licensing testing to local companies.[139]

**European Union** The EPO granted Myriad patents on the BRCA1 and BRCA2 genes in January and May of 2001, respectively. Before this decision, there had been enormous pressure from the UK and French genetics research communities, opposing the commercialisation of BRCA testing. After patents were granted, opposition has crystallised into opposition proceedings against the EU patents by the Institut Curie in Paris and a coalition of 16 other French laboratories.[140]

Myriad Genetics filed a patent application on the discovery of BRCA, including a 'composition of matter' patent on the gene itself and a 'method of use patent' for the application of BRCA1 in the diagnostic and therapeutic arena. The sheer breadth and scope of patent claims can be inferred from the abstract of the patent application, which states: 'Specifically, the present invention relates to methods and materials used to isolate and detect a human breast and ovarian cancer predisposing gene (BRCA1), some mutant alleles of which cause susceptibility to cancer, in particular, breast and ovarian cancer.'[141] The patent application envisions a number of methods of use, ranging from the preparation of recombinant or chemically synthesised nucleic acids, to nucleic acid and peptide therapy. For therapies, which rely on BRCA1, Myriad sold those rights to Indianapolis-based Eli Lilly and Co., retaining rights for sequencing BRCA1 with itself.[142]

A number of foreign equivalents to this European method for diagnosing a predisposition for breast and ovarian cancer exist. In France, the Institut Curie, a reputed medical and research institution, initiated an opposition procedure against the patent EP 699754 granted to Myriad Genetics for a method for diagnosing a

---

[137] *Association of Molecular Pathology et al. v. United States Patent and Trademark Office et al.*, available at http://www.cafc.uscourts.gov/images/stories/opinions-orders/10-1406.pdf (Visited on August 30, 2011).

[138] Sec. 23, H.R. 1249—THE "LEAHY-SMITH AMERICA INVENTS ACT" As Passed by the House on June 23, 2011, available at http://www.uspto.gov/patents/init_events/section_summary_26jul2011.pdf (last visited on May 22, 2012).

[139] *Id.* at 138.

[140] *Ibid.*

[141] Rimmer, *supra* note 13 at 191.

[142] *Ibid.*

predisposition for breast and ovarian cancer associated with the BRCA1 gene. The Institut Curie challenged the patent granted by the EPO on the grounds of lack of novelty, lack of inventive step and insufficient description. The action was supported by other institutes from France, Belgium and Italy.[143]

The opponents objected to the discrepancies in ten DNA letters between Myriad's original 1994 patent application and the BRCA1 gene sequence described in Myriad's patent issued in 2001. They emphasised that it was not until 1995 that Myriad submitted a sequence matching exactly the one in the issued patent. By that time, the crucial sequence had already been publicly open on the scientific database, GenBank, and fell under the so called 'prior art'. After hearing both the parties, the Opposition Division of the EPO revoked the patent, EP 699754.[144]

Regarding the lack of novelty allegation by the Institut Curie, the Opposition Division of the EPO looked into whether the patent application complied under Article 54 of the EPC 1973 with the requirement for novelty. As regards to the novelty criterion, the Opposition Division focussed on the issue whether a full and complete version of the genetic sequence was first disclosed by Myriad genetics Inc. or published on the genetic database, Genbank, and observed that 'there arises some doubt as to the exact nature of the BRCA1 sequences which were available from GenBank at the time when D1 and D2 were published.'[145] In the absence of sufficient evidence from the opponents to substantiate their allegations, the Opposition Division found itself unable to reach a decision on this point based on the opponent's arguments alone or to establish the facts of its own motion.[146]

The second allegation of Institut Curie was that the application lacked an inventive step: 'The patent application, as granted has an excessively broad scope which does not correspond to the significance of Myriad's contribution to the public domain, at the date the patent was filed.' As regards to the lack of inventive step, the Opposition Division of the EPO held that the patent application failed to comply with Article 56 of the EPC 1973, which requires an inventive step.[147] The Opposition Division observed that the requirement of an inventive step has a higher threshold than mere novelty because it addressed the question of whether or not an invention was inventive as compared to the prior art.[148]

Regarding the last allegation of lack of adequate description, the Opposition Division held that the patent application satisfied the requirements of Article 83 of the EPC. However, the Opposition Division did not agree with the allegations of the Institut Curie that a 'number of BRCA1 mutations are still unclassified with regard to their diagnostic significance, such that the claims of the request are not enabled across their whole breadth.'[149]

---

[143] *Id.* at 192.
[144] *Ibid.*
[145] *Ibid.*
[146] *Ibid.*
[147] *Id.* at 193.
[148] *Ibid.*
[149] *Ibid.*

4.6 Patenting of Genetic Tests for Diagnostic Purposes

A report of the New York Times showed that the stock of the Myriad Genetics dropped 1.7% in value after the European decision. As a result, Myriad Genetics Inc. quietly divested itself of the ownership of the patent, assigning its interest to the University of Utah Research foundation.[150]

The Institut Curie and a number of parties filed an opposition procedure against the patent EP 705902 on the similar grounds viz. lack of novelty, lack of inventive step and insufficient description.[151] The Opposition Division of the EPO upheld the validity of the patent in an amended form.[152] The Opposition Division held that EP 705902 had demonstrated an inventive step over the prior art.[153] It concluded that 'the BRCA1 sequence errors reported in the priority documents were caused by technical difficulties that had been overcome by repetitive sequencing.' The Division maintained: 'These errors were not the result of the use of probes which were unsuitable to isolate the BRCA1 gene.'[154]

The issue of morality was also raised by the opponents, who argued that 'the patent application offended public morality and notions of human rights, because it would hinder research and laboratory testing and it would lead to the serious obstruction of public health systems.'[155] However, the logic provided by the Opposition Division was somewhat odd. While dismissing the ethical and human rights objections of the opponents, the Opposition Division characterised them as merely economic arguments. The decision shows how the EPO narrowly construes exclusions from patentable subject matter on the basis of '*ordre* public or morality'.[156]

As regards to the validity of the patent claims of Myriad genetics and the cancer research campaign to inventions relating to BRCA2, the patent application filed by Myriad Genetics Inc. was related to the Chromosome 13-linked breast cancer susceptibility gene BRCA2, covering a wide range of diagnostic applications. The EPO granted the patent in January 2003. The Belgian Society of Human Genetics and the Institut Curie filed opposition proceeding against the patent later in December 2003. In the meantime, Myriad Genetics Inc. transferred ownership of the patent to the University of Utah Research Foundation in 2004.[157]

**Patents Were Allowed in Amended Form** After holding hearings, the Opposition Division of the EPO upheld the validity of the European patent EP 785216 in an amended form. The patent claims were limited to the use of a particular nucleic acid carrying a mutation of the BRCA2-gene which is associated with the predisposition to breast cancer for in vitro diagnosing of such a predisposition in Ashkenazi Jewish women. This decision started a fresh debate about the validity of patent claims,

---

[150] *Id*. at 194.
[151] *Ibid*.
[152] *Ibid*.
[153] *Ibid*.
[154] *Ibid*.
[155] *Ibid*.
[156] *Id*. at 195.
[157] *Id*. at 196.

focussing on identification of one particular mutation 'for diagnosing a predisposition to breast cancer in Ashkenazi Jewish women'.[158] As a result of the decision, the EPO has been criticized for countenancing racial and genetic discrimination.[159]

In 2001, the European Parliament passed the resolution on the patenting of the BRCA1 and BRCA2 genes. It 'reiterates its call on the Council, the Commission and the Member states to adopt the measures required to ensure that the human genetic code is freely available for research throughout the world and that medical applications of certain human genes are not impeded by means of monopolies based on patents'.[160]

The European Parliament had serious concerns that the granting of patents by the EPO could create a monopoly for Myriad Genetics, which could have serious implications for the further use of existing genetic tests for breast cancer. It stressed that 'this development could have an unacceptable detrimental effect on the woman concerned and constitute a serious drain on the funds of public health services; whereas moreover it could seriously impede the development of and research into new methods of diagnosis.'[161]

Despite the awareness of the ill effects of gene patents, the EPO is able to patent it, because it is not the legislature which has to balance conflicting interests and lay down legal rules. The EPO is an administrative agency which applies and interprets the rules laid down by the legislature. There remains a great division among European Union members over the European Union Directive on the Legal Protection of Biotechnological Inventions 1998 (EU).[162]

**Canadian Position** Myriad's patents on BRCA1 and BRCA2 genes have serious implications for health policies in Canada. Canada is entangled between two approaches: first, European approach to the delivery of healthcare and social programs and second, the USA's approach for economic growth. In the context of human gene patents, Canadian policymakers are struggling to control the cost of the publicly funded healthcare system. At the same time, they are also concerned with the importance of biotechnology in economic growth. Balancing these two policy goals is a great challenge before the Canadian policymakers.[163] Despite ethical objections against human gene patents, thousands of patents involving human gene have already been issued. A review of Intellectual Property Office Canadian Patents Database in 2002 found approximately 2100 patents associated with human DNA.[164]

Despite the existence of a pro-patent environment, the push to the academic research to regional economic growth is of recent origin. This is reflected in the

---

[158] *Ibid.*
[159] *Ibid.*
[160] *Ibid.*
[161] *Ibid.*
[162] *Id.* at 203.
[163] Caulfield 2005.
[164] *Id.* at 224.

## 4.6 Patenting of Genetic Tests for Diagnostic Purposes

policies of Canada's major public research funding entities such as Canadian Institute of Health Research (CIHR). CIHR now have a formal commercialisation mandate. Its enabling legislation states that the CIHR 'should encourage innovation, facilitate the commercialisation of health research in Canada and promote economic development through health research in Canada'.[165] Gene patenting in Canada remains a controversial issue because of its impact on access to healthcare services and the potential cost of providing useful healthcare services within a publicly funded system.[166]

In Canada, the decision by the US-based multinational company, Myriad Genetics, to take steps to enforce the patents on the BRCA1 and BRCA2 genes has created furore in the society. In Canada, offering the BRCA1 and BRCA2 test to women with a specific at-risk profile is viewed as part of the clinical standard of care. It is offered throughout the country as a part of the provincial healthcare schemes, albeit with some degree of variation in clinical and laboratory practice. These tests are generally conducted in one of the public research or provincial diagnostic laboratories. Myriad started sending cease and desist letter to most Canadian provincial health care ministers in 2001.[167] It was mentioned in the letters that all future genetic testing that utilises the BRCA1 and BRCA2 genes must be done through Myriad's laboratory.[168] Various Canadian provinces funded their own tests for detecting mutations in the breast cancer genes until they received cease and desist orders from Myriad in 2001. These orders claim that these provinces were infringing Myriad's patents.[169] The argument by the provinces that the test used by them was different from the one claimed by Myriad in its patents was not found relevant because ultimately any screening test required use of patented genes.[170] The cease and desist letters required that DNA samples from Canada to be sent to Myriad's laboratory for testing. The letters cautioned the provinces that if they failed to comply they might face the risk of patent infringement litigation.[171] The Myriad test which costs approximately Canadian $ 38,000, is, in some cases, more than four times the cost of the testing being done within the provincial system.[172] This has raised concerns among the health ministers about the heavy cost burden on their respective provincial healthcare system.[173]

In Canada, there have been diverse approaches adopted by different provinces. British Columbia initially suspended its funding for the test, then withdrew funding altogether, and has since resumed funding for a different test; Alberta and Manitoba continued as they had been doing; Saskatchewan, Newfoundland and Nova Scotia

---

[165] *Ibid.*
[166] *Ibid.*
[167] *Ibid.*
[168] *Ibid.*
[169] Garforth 2005
[170] *Id.* at 80.
[171] *Ibid.*
[172] tim
[173] tim

sent their samples to Ontario, so dependent on the latter's decision; Quebec complied with the order and began sending its samples to Utah. In September 2001, Ontario announced that it would not comply with the order and would continue to fund its own tests believing that it was not infringing Myriad's patent. Afterwards, Ontario adopted a new test that is cheaper and more accurate than that the one it had been using previously. No action has been taken by Myriad against the practice of Ontario till now.[174]

The public healthcare system funded by the Ontario government faced a difficult situation in the late 1990s when the test was still considered experimental in Canada. The test was available to Canadian women only if they participated in the research studies, and these studies could take up to 2 years to return with the test results. On the other hand, Myriad's results were available within a few weeks. An Ontario woman named Fiona Webster had a family history of breast cancer and she wanted to know if she was at risk or not. She was 39 at that stage and felt an urgent need of a test result without waiting for two long years. The Ontario Health Insurance Plan (OHIP) was ready to pay $ 20,000 to Webster to have a bilateral mastectomy but was reluctant to pay for Myriad's test. Webster appealed the OHIP decision to the Ontario Health Services Appeal Board which ordered that the government must pay for the Myriad test if an individual can show a compelling case history. It was found that Webster did not have mutation in her genes, so the surgery would have been unnecessary. This incident reflects the vulnerability of the provincial governments to face a similar litigation if they refuse to fund a genetic test.[175]

In response to the Myriad controversy, the Ontario government struck a policy group to examine the potential adverse social implications of gene patenting. The 2002 Report of the Premiers of the Ontario government recommends a clarification of patent criteria in relation to human genes, the exclusion of broad-based genetic patents covering multiple uses, a clarification of the experimental and non-commercial exceptions and an expansion of the methods of medical treatment exclusion.[176] It further recommended for the introduction of a compulsory licensing scheme in order to ensure access within the public access system.[177]

Myriad's patent over BRCA genes has serious implications in Canada. The implied or real threat of patent infringement may delay or block the development, validation and implementation of diagnostic tests by Canadian laboratories. The method mandated by the patent holder for conducting the test may not be the most appropriate for the patient. The high price charged by patent holders for genetic tests may cause provincial health care systems to refuse to insure these tests. The high costs of tests not covered by provincial health insurance plans may render these tests unaffordable and unavailable to many patients. Sending patient samples out of Canada to a company not subject to Canadian laws and regulations may cause ethical concerns over quality control and confidentiality.[178] In Canada, physicians

---

[174] Garforth, *supra* note 66 at 80.

[175] *Id.* at 81.

[176] *Tim* at 226.

[177] *Ibid.*

[178] Gold et al. 2002.

and their patients face trouble in having the appropriate test. Physicians are duty bound to advice their patient to have a clinically useful test in order to provide them successful disease management.[179] However, they have no control over the quality of these tests.

## 4.7 Arguments in Favour of Patents on Diagnostic Tests—to Develop Diagnostic Tests Require Significant Efforts

Some argue that since diagnostic genetic testing is a new area, therefore, patents are required for its development. In the absence of patent protection, the invention and development of new diagnostic tests would be seriously hampered. Despite the fact that the human genome has been sequenced, locating a particular gene does not itself lead directly to a test being available. It is true that developing a genetic test, once the gene associated with an inherited disease has been identified, is sometimes a routine and relatively straightforward task. However, the scenario may not be the same when the testing of very large genes, multiple mutations, or multiple genes or fragments of genes is required.[180] It may need 'significant effort to convert the basic knowledge of genetic structure into a clinically applicable, reliable, diagnostic test, although the investment required is unlikely to approach that needed for bringing a medicine through the regulatory process to market.'[181] The situation warrants some sort of incentive structure in the form of a patent to develop these diagnostic tests.[182] Nuffield Council, however, does not borrow this argument. It commented:

> We are not persuaded by this argument. We note that there are other approaches which could, in the future, offer inexpensive, rapid predictive knowledge relating simultaneously to more than one disorder, based on the straightforward and valid use of patented technologies, rather than what is essentially patented information. For example, the development and application of patented technologies such as DNA microarrays have considerable potential for the precise diagnostic classification of cancer. These devices contain thousands of DNA sequences in an ordered array, which allows simultaneous analysis of a similar number of genes or marker regions of DNA that are closely associated with genes.[183]

## 4.8 Policy Implications of Myriad's Patents on BRCA1 and BRCA2

The study regarding the Myriad's patents on BRCA1 and BRCA 2 genes in the USA, European Union and Canada reveals that these patents have numerous policy implications.

---

[179] *Id*. at 257.
[180] Id. At 5.18
[181] 5.18
[182] *Id*. para. 5.18.
[183] Para. 5.21.

*Low Predictive Value* BRCA test only provides general estimated chances of developing breast cancer because of various other factors such as environmental factors that are also contributing in the disease. It has a very low predictive power for women without a family history of breast cancer, meaning thereby, that many women who test positive for BRCA1 mutation would not ever manifest symptoms of the disease.[184]

*Quality of Testing* Apart from low predictive power to test the general population for the susceptibility of breast cancer, the Myriad BRCA test reportedly fails to detect 10–20 % of all expected mutations. 'Scientists with the Institut Curie have discovered the deletion in the BRCA1 gene accounting for the predisposition to breast cancer of one U.S. family that the Myriad test fails to detect altogether.' The failure to detect such a large percentage of mutations may have an adverse impact on the quality of the test. As a result of such a failure Myriad tests may fall well short of appropriate patient care, preventing patients from availing alternative more effective tests.[185]

*Lack of Follow-up Genetic Counselling and Clinical Healthcare* In European countries, healthcare workers follow a model that integrates biological research, clinical investigation, and patient care, especially considering the psychological aspects of diagnostics, both for the individual patient and patient's family. On the contrary, Myriad is accused of dissociating genetic testing from patient care as it provides BRCA test results without any significant follow-up individualised genetic counselling. This may seriously affect the patients, impeding the quality of patient care.[186]

*Direct to Consumer Campaign Bypass Physicians* Myriad's direct to consumer advertising campaign for the BRCA diagnostic tests bypasses physicians and directly targets the general population.[187] While Myriad has assured that they will train doctors to guide patients through the process, many insist that counselling should be 'undertaken only by genetic specialists without commercial interests tied to the corporation.'[188]

*Access and Cost of Patents* Allowing patents on genetic diagnostics may create a possibility that the test would not be available to a patient for purposes of diagnosing the health problem, hindering patient care and counters the goals of the healthcare system.[189] In the case of diagnostic gene patents, doctors must either obtain a license to provide such a test or charge a patient a fee for sending a sample to be tested at the corporation or research institution that holds the patent. In many situations, this fee can be exorbitant. Due to high licensing fee, the doctor may even

---

[184] Jordan Paradise, *supra* note 11 at 147.
[185] *Ibid*.
[186] *Id*. at 147–148.
[187] Id. At 148 Jor
[188] *Id*. at 148.
[189] *Id*. at 148–149.

choose to perform an inferior procedure, perhaps resulting in inaccurate results or even failure to screen for the specific disease.[190]

*Impact on Research* Myriad's patents hardly leave any space for improving the test because researchers and physicians are most often completely barred from using any gene or protein sequences claimed within the patent, and thus are prevented from undertaking or improving the diagnostic technology relating to the particular gene. This may have a deleterious effect on innovation and future research, which may ultimately result in an intellectual standstill. Because researchers and physicians are barred from the use of the BRCA1 gene itself, no improvements in the inaccuracies of the current testing mechanisms will be discovered.[191]

*Monopoly over the Genetic Material and Samples* The mandatory export of all tissue samples to Myriad for testing in the USA enables Myriad to build up the only BRCA databank in the world, giving Myriad total control over the key research materials relating to genes coding for breast cancer susceptibility. With such a monopoly over the biological material in the form of exclusive tissue bank, Myriad can make further discoveries and file patent applications to the exclusion of all other nations and researchers.[192]

## 4.9 The Possible Way Outs

*Criteria of Patenting Should be Applied Stringently or Amended* In order to ensure the accessibility of genetic tests and promote genetic research, the criteria for patenting should be stringently applied or amended. Nuffield Council opined that 'patent offices should critically assess whether the isolation of DNA sequences, in particular human DNA sequences, can any longer be viewed as inventive.'[193] It feels that in the majority of cases, this criterion will not be met.[194] The council recommended that 'the criteria already in place within existing patent systems for the granting of patents, particularly the criterion of inventiveness be stringently applied to applications for product patents which assert, inter alia, rights over DNA sequences for use in diagnosis.'[195] It further recommended:

> ...[t]he European Patent Office (EPO), the United States Patent and Trademark Office (USPTO) and the Japan Patent Office (JPO) together examine ways in which this may be achieved. If this recommendation were to be implemented, we expect that the granting of product patents which assert rights over DNA sequences for use in diagnosis would become the rare exception, rather than the norm. Where the application of the criterion of

---

[190] *Id.* at 149.
[191] *Ibid.*
[192] *Id.* at 150.
[193] *Ibid.*
[194] *Ibid.*
[195] *Ibid.*

inventiveness is not particularly stringent, as for example in the US, additional mechanisms may be needed. We recommend, accordingly, that the USPTO and US lawmakers give consideration to whether patent laws need to be amended for this purpose.[196]

*Product Patents Should Be Discouraged* A product patent relating to a DNA sequence provides the patent owner exclusive rights to all subsequent uses of that sequence.[197] 'A product patent on a diagnostic test for a gene would allow the patent owner a monopoly on all uses of that sequence for any sort of test or other application. For example, if BRCA1 were found to be linked to heart disease, or cancer of the bladder, the rights of the owner of the product patent would extend to these new diagnostic tests or other applications.'[198] One possible option in this regard is to limit patents on diagnostic tests based on DNA sequences to use patents, that is, patents which do not assert rights over the DNA sequence itself.[199]

*Patents Should Be Restricted to Specific Use* Nuffield Council concludes that 'the protection by use patents of specific diagnostic tests which are based on DNA sequences could provide an effective means of rewarding the inventor while providing an incentive for others to develop alternative tests.'[200]

*A Balance Is Needed to Struck While Invoking Compulsory Licensing* Compulsory licensing is invoked in those exceptional situations, in which, the existence of a monopoly is creating an unacceptable and unfair situation.[201] Nuffield Council made the commitment:

> We do not, therefore, support a wholesale and indiscriminate use of compulsory licensing. Rather, in those specific cases in which the enjoyment of exclusive rights to the diagnostic use of a DNA sequence is not in the public interest, we recommend that those seeking to use the diagnostic tool or develop an alternative should seek a compulsory licence from the relevant authorities if they are refused a licence from the owner of those rights on reasonable terms, and we encourage the authorities to grant such a licence. We also note the suggestion made by the Organisation for Economic Co-operation and Development (OECD) of a "clearing house" to ease the obtaining of licences for "genetic inventions" by commercial laboratories. We suggest that this concept, which might reduce transaction costs, should be explored further.[202]

The debate relating to patenting of a DNA sequence has been mostly confined to the developed world. In developing countries, the patenting of drugs generated much heated debates as compared to that of genes. Developing countries are largely importers rather than producers or manipulators of genetic information, baring few exceptions such as China, which played a substantial role in the public Human Genome Project, Brazil and India. However, one of the serious concerns relating

---

[196] *Ibid.*
[197] 5.23
[198] *Ibid.*
[199] *Id.* para. 5.23
[200] *Ibid.*
[201] *Id.* para. 5.29.
[202] *Ibid.*

to developing countries is that companies could be in an even stronger position in developing countries to use their patent rights to charge extortionate prices for diagnostic tests.[203] It would be interesting to see how developing countries would carve out strategies to cope up with the challenges posed by gene patenting.

# References

Caulfield Timothy (2005) Policy Conflicts: Gene Patents and Health Care in Canada. Community Genetics 8:223.
Cornish W. (1998) Experimental use of Patented Inventions. IIC 29:735.
Duffy John F. (2002) Patent System Reform: Harmony and Diversity in Global Patent Law. Berkeley Technology Law Journal 17:685
Dutfield Graham (2006) DNA patenting: implications for public health research. Bulletin of World Health Organisation 84: 389.
Garforth Kathryn (2005) Health Care and Access to Patented Technologies. Health Law Journal 13: 77–97.
Gold E. Richard, Joly Yann, Caulfield Tim (2005) Genetic Research Tool, The Research Exception and Open Science. GenEdit 3: 1.
Heller Michael A. and Rebecca S. Eisenberg (1998) Can Patents Deter Innovation? The Anticommons in Biomedical Research. Science 280: 698
Mueller Janice M. (2001) No "Dilettante Affair": Rethinking the Experimental use exception to patent infringement for biomedical research tools. WashingtonLaw Review 76: 1
Nuffield Council of Bioethics (2002) The ethics of patenting DNA, 5.24 http://www.nuffieldbioethics.org/sites/default/files/The%20ethics%20of%20patenting%20DNA%20a%20discussion%20paper.pdf. Accessed 18 May 2011.
Paradise Jordan (2004) European Opposition to Exclusive Control Over Predictive Breast Cancer Testing and the Inherent implications for U.S. Patent law and Public Policy: A Case Study of the Myriad genetics' BRCA 1 Patent Controversy. FOOD AND DRUG LAW JOURNAL 59: 135
Patenting and the Research Exemption http://www.transknowlia.org/content/patenting-and-research-exemption. Accessed 12 August 2012
Richard Gold, Timothy A. Caulfield, Ray Peter N. (2002) Gene patents and the standard of care. Canadian Medical Association Journal 167: 256.
Rimmer Matthew (2008) Intellectual Property and Biotechnology: Biological Inventions. Edward Elgar Publishing, Inc., Massachsetts p. 187.
Safrin Sabrina (2004) Hyperownership in a Time of Biotechnological Promise: The International Conflict to Control the Building Blocks of Life. The American Journal of International Law 98: 669
Silverstein Tina, Joly Yann, Harmsen E. et al (2009) The Commercialisation of Genomic Academic Research: Conflicting Trend In: E. Richard Gold & Bartha Maria Knoppers (eds.) Biotechnology IP & Ethics LexisNexis Canada Inc., Markham, Ontario, p. 133.
Sumikura Koichi (2009) Intellectual property rights policy for gene-related inventions-toward optimum balance between public and private ownership. In: David Castle (Ed.) The Role of Intellectual Property Rights in Biotechnology Innovation. Edward Elgar Publishing Limited, Cheltenham U.K./Massachusetts U.S.A. p. 77
Thumm Nikolaus (2005) Patents for genetic inventions: a tool to promote technological advance or limitation for upstream inventions? *Technovation* 25:1410

---

[203] Dutfield 2006b.

Vicky Clark, Pitfalls in drafting royalty provisions in patent licences. Bio-Science Law Review http://www.pharmalicensing.com/public/articles/view/1087832097_40d70021d738c. Accessed 10 May 2012

Walsh John P., Arora Ashish, Cohen Wesley M. (2003) Working Through the Patent Problem. SCIENCE 299: 14

Williams Bryn (2002) History of a Gene Patent: Tracing the Development and Application of Commercial BRCA Testing. Health Law Journal 10:137.

Zekos Georgios I. (2005) Discrepancies in Biotechnology/Chemical Patenting. *Journal of Intellectual Property Rights* 10: 1

# Chapter 5
# Intellectual Property Protection to Bioinformatics and Genomic Databases and Open Source Analogy to Biotechnology

The recent advances in the field of biotechnology, particularly the development of human genomics and the success of Human Genome Project, make a gene more important because of its informational content rather than its material qualities (physical attributes). The vast amount of genetic and genomic information unleashed by biotechnological advances necessitated devising methods to manage, arrange and catalogue it in a manner that may facilitate its use. This has led to the emergence of a fairly new discipline, bioinformatics, and development of genetic and genomic databases. Intellectual property protection varies according to the technology used in the bioinformatics field such as biological databases, algorithms, complex software etc. While some of these technologies may fit into the existing framework of intellectual property law, others fall outside the scope of current legal protections e.g. databases per se, as collections or arrangements of raw data,, are generally not patentable. Bioinformatics and genomic databases as new fields need a continuously open and collaborative process for data collection and analysis. In such a situation, strict proprietary protection to genetic and genomic databases may restrict public access to genetic and genomic information. These important aspects of bioinformatics are in direct conflict with the proprietary protection given to it. The most viable option to meet this situation is seen in the open source biotechnology, inspired by the success of open source movement in the field of information technology.

## 5.1 Transition in Biotechnology: From Lab-based Technology to Computer-based Science

The rapid advances in molecular biology and genetics such as the Human Genome Project unleashed a great deal of genetic information of immense medical and therapeutic value. This necessitated devising methods to manage all the information gleaned so far and to arrange and catalogue them in a manner that may facilitate their use. Sometimes the actual utility of most information was unknown, but needed to be preserved for future use. At this point, the use of computer technology to store and catalogue the data began. Initially, the role of computer science in the molecular biology field was confined to the extent of data management and cataloguing; however, eventually the need to access information from these databases led to the development of mining software.[1] These developments led to the creation of bioinformatics, which depicts a marriage between life sciences and computer technology.[2] The information gleaned from experiments may be useful in discovery of cures for various diseases and also help understand reasons for incidence of diseases at the genetic level.[3]

The emergence of bioinformatics has led to a radical transformation in the field of biotechnology as laboratory-based biotechnology has been transformed into computer based science. This new field of biotechnology focuses upon automated collection, compilation, storage, retrieval and analysis of biological data. This led to discoveries and innovations far beyond the scope of conventional biotechnology.[4]

### 5.1.1 Definition of Bioinformatics

The National Centre for Biotechnology Information defines it as a field of study in which biology, information technology and computer science merge together to form a single discipline.[5] It is also defined as the use of computer programs and other attributes of computer science to manage, catalogue and access the vast realm of biological information available.[6] Bioinformatics involves technology, which comprises mostly programs and software which would aid in the compilation and updating of extensive databases of information, databases themselves, and also includes software which aids in the retrieval, analysis and comparison of relevant data.[7]

### 5.1.2 Objection to the Extension

Objections against the extension of IPR protection to the field of bioinformatics are being made on the ground that such protection would act to enclose the ethically sensitive realm of human gene-related studies.[8] It is contended that human genomic science should be common and accessible to all and should not be restricted to few individuals. It is also contended that since the underlying purpose behind the bioinformatics is to further medical treatment, therefore, patent protection should not be

---

[1] Gopalan (2009).

[2] *Ibid.*

[3] *Id.* at 47.

[4] Hultquist et al. (2003).

[5] Just the facts: a basic introduction to the science underlying NCBI resources. Available at http://www.ncbi.nlm.nih.gov/About/primer/bioinformatics.html in *supra* note 1, at 46.

[6] Gopalan, *supra* note 1, at 46.

[7] *Id.* at 48.

[8] *Ibid.*

extended to this field.⁹ However, intellectual property advocates argue in favour of extending the patent protection to bioinformatics on the basis that such protection is necessary for encouraging innovation in this new field. Intellectual property protection to bioinformatics can be divided into two categories: bioinformatics databases and bioinformatics software.[10]

## 5.2 Bioinformatics Databases

Bioinformatics database is a large, organised body of persistent data, usually associated with computerised software designed to update, query and retrieve components of the data stored in the system. These databases are aimed at providing easy access to information and facilitate retrieval of data for analysis and comparative studies.[11]

### 5.2.1 Intellectual Property Protection to Bioinformatics

#### 5.2.1.1 Patentability of Bioinformatics Database

Bioinformatics database is a mere composition of information of abstract nature, which describes composition of DNA and RNA molecules. As a composition of abstract information, it does not constitute patentable subject matter.[12] However, patent protection may be extended to a bioinformatics database 'if it is not a mere catalogue, but is more along the lines of data processing system that has the ability to convert the raw data into a tangible result.'[13] The US Court of Appeals for Federal Circuit (CAFC) held that a data processing system is patentable subject matter as it involves the practical application of a mathematical algorithm, formula or calculation leading to a useful, concrete and tangible result.[14] Applying this interpretation to the bioinformatics databases, it is found that they are not merely compilation of data but data processing systems and may fall under the category of patentable inventions.[15]

**USA Position** In the USA, the patent law excludes from patentability abstract ideas, scientific laws, naturally occurring phenomena, products of nature, mental steps and printed matter. However, it has been the practice of the US Patent and

---

[9] *Ibid.*

[10] Just the facts: a basic introduction to the science underlying NCBI resources. Available at http://www.ncbi.nlm.nih.gov/About/primer/bioinformatics.html cited in Gopalan, *supra* note 1 at 48.

[11] Gopalan *supra* note 1 at 48.

[12] *Ibid.*

[13] *Ibid.*

[14] *Id.* at 49, *State Street Bank v. Signature Financial Group* 149 F 3d 1368.

[15] *Ibid.*

Trademark Office (USPTO) to interpret cases such as *State Street Bank and Trust v. Signature Financial Group*[16] and *ATT v. Excel Communications*[17] to make bioinformatics software patentable. In order to make software patentable, the data residing in the software must 'interact' with a computer readable medium i.e. they must be able to direct a computer to accomplish a particular result.[18]

Since bioinformatics innovations primarily involve computer-based applications such as database and software for the collection and processing of biological data, therefore, in the USA, the USPTO defines this category as 'inventions implemented in a computer-readable media.'[19]

Computer-based inventions fall under two heads: hardware aspect and software aspect.[20] The hardware aspect of computer-based inventions is protected under US patent law like any other mechanical invention. On the other hand, the software aspect of computer-based inventions has been the subject of intense debate for decades. Till 1970, US courts and the USPTO rejected patent claims on computer software, considering software as printed matter, mental steps, algorithms or abstract ideas which fall under the patent exclusions.[21] However, this position has been challenged in successive cases.[22] These legal challenges compelled USPTO to reconsider its previous rejection of patent claims. Accordingly, in 1995, after IBM appealed the USPTO's rejection of a claim to 'software continued on a floppy disk' before the US CAFC, the USPTO withdrew the opposition to IBM's claim and, in January 1996, issued the final version of the Examination Guidelines for Computer-Related Inventions ('the Guidelines').[23]

While complying with the decisions by the CAFC in the cases, *In re Warmerdam*[24] and *In re Lowry,*[25] the USPTO explained the reason for the statutory distinction between data structure per se and those encoded on a computer-readable medium or machine in the Guidelines:

> Data structures that are not claimed as embodied in computer-readable media are descriptive material per se and are not statutory because they are neither physical "things" or statutory processes. Such claimed data structures do not define any structural and functional inter-relationships between the data structure and to her claimed aspects of the invention which permit the data structure's functionality to be realised. In contrast, a claimed com-

---

[16] State *Street Bank v. Signature Financial Group* 149 F 3d 1368.

[17] *AT & T Corp. v. Excel Communications*, Inc. 172 F.3d 1352 (1999).

[18] Rees (2003).

[19] Hultquist and others, *supra* note 4; U.S. Patent and Trademark Office. Examination guidelines for computer-related inventions. 61 CFR 7478.

[20] *Ibid.*

[21] *Ibid.*; In re Prater, 415 F.2d 1378 (CCPA 1968), *Gottschalk v. Benson*, 409 US 63 (1972).

[22] *Ibid.*; Parker v. Flook, 437 US 584 (1978), *Diamond v. Diehr*, 450 US 175 (1981), In re Freeman, 573 F.2d 1237 (CCPA 1978), In re Walter, 618 F.2d 758 (CCPA 1980), In re Abele, 684 F.2d 902 (CCPA 1982), In re Alappat, 33 F.3d 1526 (CCPA 1994), *State Street Bank & Trust Co. v. Signature Fin. Group, Inc.*, 149 F.3d 1368 (Fed. Cir. 1998).

[23] *Ibid.*; In re Beauregard, 53 F.3d 1583 (Fed. Cir. 1995).

[24] In re Warmerdam, 31 USPQ 2d 1754 (Fed. Cir. 1994).

[25] In re Lowry, 32 USPQ 2d 1031 (Fed. Cir. 1994).

puter-related medium encoded with a data structure defines structural and functional interrelationships between the data structure and the medium which permit the data structure's functionality to be realised, and it is thus statutory.[26]

Although the Guidelines make a difference between a data-structure per se and a computer or computer-readable medium encoded with a data structure, however, in practice, since the computer or the computer-readable medium does not have to be limited to any specific embodiment, the recital in the patent claims of these items imposes no impediments to meeting the 'structural and functional inter-relationships' requirement of the Guidelines.[27]

**European Union** Like the USA, the patenting of software has been problematic in Europe also. The European Patent Convention (EPC) excludes computer programs from patentability. However, the European Patent Office (EPO) in Germany realised very soon after its foundation that this exclusion was illogical.[28] In VICOM decision,[29] the EPO pointed out that the wording of the EPC excluded only the patenting of computer programmes as such.[30] Nevertheless, a general-purpose computer programmed for a special purpose is not excluded from patentability as long as it produces a technical effect.[31] The VICOM decision opened the way for the patenting of inventions implemented by means of computers in Europe.[32]

*Protection of Bioinformatics Databases in the European Union* In the realm of bioinformatics, initially, databases had been stored as a flat file structure, however, in due course of time, more sophisticated relational databases were developed. These sophisticated relational database structures were developed to allow more efficient and significant analysis of the data stored therein.[33] Information contained in databases can either be protected through copyright or sui generis database rights. Here, the extent to which database information can be protected by copyright varies widely depending on the country involved. In most of the countries, copyright protection is not available for information contained in databases.[34] However, countries, such as Australia,[35] consider that the arrangement and collection of the information may be so significant that copyright can be granted on the database.[36]

---

[26] Patent and Trademark Office United States Department of Commerce. *Examination guidelines for computer-related inventions.* Available at http://www.uspto.gov/web/offices/pac/dapp/pdf/ciig.pdf (last visited on 12 May 2012).

[27] Hultquist and others, *supra* note 4.

[28] *Ibid.*

[29] T 0208/84 of 15 July 1986 Computer-related invention/VICOM. *Official J. Eur. Pat. Off.* 14–23 (1987).

[30] Hultquist and others, *supra* note 4.

[31] *Ibid.*

[32] *Ibid.*

[33] Hultquist et al. (2002).

[34] *Ibid.*

[35] *Telstra Corporation Limited v. Desktop Marketing Systems Pty Ltd*, [2001] FXA 612 (15 May 2001).

[36] Hultquist et al., *supra* note 33.

*European Database Rights Directive* European Union adopted the European Database Rights Directive to harmonise protection within member states. The directive protects 'a collection of independent works, data or other materials arranged in a systematic or methodical way and individually accessible by electronic or other means'.[37] Accordingly, a developer of a database can prevent the extraction and/or re-use of the whole or a substantial part of the contents of the database.[38] However, the protection provided under the directive is limited only to persons or legal entities residing in the European Economic Area (the European Union, Norway, Iceland and Liechtenstein) or in countries having similar protection schemes. The USA has not endorsed proposals to introduce a similar patent.[39]

Although patent protection cannot be extended to the informational content of the database, however, such protection can be extended to the structure of the database.[40] While considering the patentability of a data structure[41] in a patent application for a picture retrieval system having data stored on or in a record carrier of a particular structure, the board maintained that 'there was a difference between the functional data, which controlled the technical working of the system, and the cognitive information, which represented the picture that could be retrieved and displayed'[42] As data relating to the DNA sequences or protein structure are not merely 'cognitive information,' it is possible to argue successfully that data structures containing this information are patentable.[43]

*Patenting of Algorithms* Due to considerable growth of data in academic and commercial databases, it becomes pertinent to develop algorithms such as Smith–Waterman algorithm.[44] Algorithms per se are not patentable in Europe; however, patent protection can be secured for it when it involves a practical application.[45] The EPO explains in its Guidelines for Examination[46] that an electrical filter designed using a mathematical method would not be excluded from patentability.[47] While dealing with the patentability of the interactive rotation of displayed graphic objects on a screen,[48] the EPO Board of Appeal stated that the invention did not relate to a mathematical method as such, but that the 'calculating steps mentioned are only

---

[37] Directive 96/9/EC of the European Parliament and of the Council adopted on 11 March 1996 on the legal protection of databases. O.J. Eur. Union No. L77, 27 March 1996, 20 (the 'Database Directive'), Art. 1, para. 2

[38] *Id.* Art. 7, para.1.

[39] Hultquist et al. *supra* note 33.

[40] *Ibid.*

[41] Decision T 1194/97–3.5.2 of 15 March 2001. Data Structure Product/PHILIPS.

[42] *Id.* Point 3.3.

[43] Hultquist et al. *supra* note 33 at 518.

[44] *Ibid.*

[45] *Ibid.* (Though, the lines of code that implement the algorithm are protectable by copyright.).

[46] Part C, Chapter IV, 2.3.3, European Examination Guidelines, available at http://legis.obi.gr/espacedvd/legal_texts/gui_lines/e/c_iv_2_3_3.htm (last visited on 12 May 2012).

[47] Hultquist et al. *supra* note 33 at 518.

[48] T 0059/93–3.5.1. of 20 April 1994.

## 5.2 Bioinformatics Databases

means, or tools, used within the overall method claimed, for entering a rotation angle value into a draw graphic system.'[49] This clearly suggests that an algorithm used in the analysis of the DNA sequence or protein data should be patentable, if it is not couched in purely mathematical terms but is applied to achievement of a useful, concrete and tangible result.[50] Accordingly, an algorithm used in identifying homologies among genes should be patentable because it offers a useful, concrete and tangible result, and is only a means of obtaining information about the homologies. Likewise, an algorithm to mine existing data for potentially useful properties is also protectable.[51]

*Interfaces* In a 1988 decision[52], the EPO considered the patentability of a user interface. In this case, a method was claimed for displaying one of a set of predetermined messages indicating a specific event that may occur in an input/output device of a word-processing system.[53] The EPO Board of Appeal maintained that giving visual indications automatically about conditions prevailing in an apparatus or system is basically a technical problem[54] and, thus, is not excluded from patentability. It is very likely that EPO would favour generally the patentability of an interface through which information is exchanged about conditions prevailing in an apparatus or system.[55]

**Indian Law** 'Such a manipulation is possible under Indian law which does not allow patents in mere presentation of information or computer program *per se*.'[56] A bioinformatics database can be a suitable candidate for patent protection because 'it is neither mere presentation of information nor a mere computer program, but both combined with other operations which can be used in a number of applications.'[57]

### 5.2.1.2 Viability of Patent Protection with Respect to Bioinformatics Databases

*Only the Method and not the Content Is Protected Through Patents* A patent for bioinformatics database is merely on the process of compiling and operating the database, namely software or computer program. It does not extend to the data within the database, which still remains mere information and is, therefore, not pat-

---

[49] *Id.* Point 3.2.

[50] Hultquist et al. *supra* note 33 at 518.

[51] *Ibid.*

[52] T0115/85 of 5 September 1988 Computer-related invention/IBM. Official J. Eur. Pat. Off. 30–34 (1990).

[53] Hultquist et al. *supra* note 33 at 518.

[54] *Supra* note 52 Point 7.

[55] Hultquist et al. *supra* note 33 at 518.

[56] Gopalan, *supra* note 1 at 49; See Sec. 3 of the Patents Act 1970.

[57] *Ibid.*

entable subject matter. Therefore, protection accorded to bioinformatics database by patent law is a mere token protection and does not ensure total exclusivity of the data compiled. An infringer may still utilise the compiled data to create an independent database by merely modifying the algorithm or software used to generate the database.[58]

*Prior Art Problem* Compilation of bioinformatics databases has been happening since the genesis of the field of study in early 1980s. Therefore, there might be problems in receiving a patent also on the grounds of prior art and non-obviousness of the claim.

### 5.2.1.3 Copyright Protection to Bioinformatics Database

Copyright seems to be the most efficacious mode of protection for the bioinformatics databases. In the USA, law on copyright protection could be extended to compilation. The US Supreme Court, when interpreting the position on copyright protection accorded to compilations,[59] held that facts are not copyrightable but compilation of facts are, provided there is sufficient degree of originality in the compilation in terms of selection and arrangement of terms, in terms of indices employed etc. However, under this law, copyright protection is extended only to the compilation to the extent of its original selection and arrangement and not to the contents.[60]

### 5.2.1.4 EU Directive on Protection of Databases

The European Union Directive on protection of databases provides for the protection of the contents of the database coupled with the protection for the database if there is originality in the selection and arrangement of material. The EU Directive based on the rationale that a person who has made a substantial investment in obtaining, verifying or presenting the database must have right of exclusivity over it. The directive protects against unauthorised extraction of the information or utilisation of the whole or a substantial part of the database.[61]

### 5.2.1.5 A Combination of Copyright Protection and Database Rights

Some suggest that the best form of protection which can be accorded to the bioinformatics database would be through copyright law on compilation coupled with the rights guaranteed under the EU Directive. Such a combination would protect compilation to the extent of its selection and arrangement and also the contents from extraction and re-utilisation. The traditional rights, such as right to reproduction,

---

[58] *Ibid.*
[59] *Fiest Publications Inc v. Rural Telephone Service Co.* 499 U.S. 340.
[60] Gopalan, *supra* note 1 at 49.
[61] *Ibid.*

licensing and publication would still apply and would be vested in the person making the compilation. The fact that most of these compilations are computer generated documents, it is generally agreed that the end product is an expression of author's idea through the medium of computer.[62]

## 5.3 Bioinformatics Software

### 5.3.1 Patent Protection to Software

Software now constitutes patentable subject matter under the US law, if it produces a useful, concrete and tangible result. The Supreme Court and Federal Circuit have indicated that so long as a software program is more than a mere algorithm, the program may be eligible for patent protection. Bioinformatics software is therefore eligible for the same protection as the software can be used for the purpose of biological research to produce results which are tangible, concrete and useful. The results acquired from the data analyses utilising these software applications have wide ranges of uses in medical diagnoses, to design drugs or draw evolutionary conclusions.[63]

Under Indian law, there is no patent available for a computer program per se.[64] However, the term *'per se'* is open to interpretation and a computer program coupled with some hardware component may fall under the scope of patentable subject matter, provided the claim is cleverly constructed in such a manner that the patent appears to be for the hardware, but the protection is claimed for the software as well, as an integral component.[65]

There is also a trend to build or manufacture customised hardware components with specific bioinformatics software keyed in. Such hardware apparatuses will fall under the scope of patentable subject matter as a machine or apparatus.[66]

### 5.3.2 Copyright Protection to Software

In the realm of copyright, the term literary work has been construed to include software and protection accorded to software under copyright has been extended to human identifiable language, source code and machine readable component and object code.[67]

---

[62] Sec. 2 (o) of the Copyright Act, 1957 which includes computer databases under the definition of literary works. Cited in *Supra* note 1 at 49.
[63] Gopalan, *supra* note 1 at 50.
[64] Sec. 3 (k) of the Patents Act, 2005.
[65] Gopalan, *supra* note 1 at 50.
[66] *Ibid.*
[67] *Ibid.*

Under the Indian law, software is included in the definition of literary work.[68] Also computer program has been defined to include both source code and object code.[69] Protection is extended to computer program as long as the work is an original expression of the idea of the person creating the program.[70]

**Copyright Is Not Effective** Copyright protection for computer software is not the best alternative, as the protection is same as extended to a literary work, and, therefore, extends only to the original expression of the idea. Therefore, it is eminently possible for a person to merely change some aspect of the object and source code to claim an independent copyright, as long as it does not become a substantial copy of the original.[71]

## 5.3.3 Trade Secret

The definition of trade secret includes software and, therefore, bioinformatics software can be protected through the medium of trade secrets. It is a usual practice for code writers to maintain the source code of their programs as a trade secret, releasing only the object code for sale or license. However, in the realm of bioinformatics software, where there is a definite desire to market the product, there is possibility that the trade secret may be disclosed by reverse engineering. The object code may be used to reach the source code, and once this is done, protection effectively collapses. The danger of reverse engineering also applies to customised bioinformatics, apparatus which might be stripped down and each individual component analysed to understand the protected trade secret.[72] There remains great confusion as to the intellectual property protection relating to bioinformatics as each form of protection viz. patents, copyrights and trademarks, has some inherent limitations. Since bioinformatics is a relatively new area, things will become clearer as the field matures.

## 5.4 Intellectual Property Protection to Genomic Databases and Problem of Accessibility

The completion of human genome sequences promises great advancement in the field of medical science. This promise can only be fully realised if researchers have full access to information gleaned from this landmark effort and subsequent research

---

[68] Sec. 2 (o) of the Copyright Act, 1957 which includes computer databases under the definition of literary works. Cited in *Supra* note 1 at 50.
[69] Sec. 2 (ffc) of the Copyright Act, 1957.
[70] Gopalan, *supra* note 1 at 50.
[71] *Ibid.*
[72] *Ibid.*

## 5.4 Intellectual Property Protection to Genomic Databases and Problem of Accessibility

initiatives. However, since biotechnology has also created enormous commercial possibilities, many private companies seek to limit access to this information. There are two conflicting views regarding the privatisation of genetic information. Some argue that the privatisation will impede scientific information, while others maintain that privatisation of genetic information is necessary to ensure profits and generate considerable amount of funds to bring therapeutic products to the market.[73]

### 5.4.1 Goals of Genomic Databases

Attempts have been made to make human genome sequence information freely accessible to researcher since 1970. For instance, Human Genome Initiative has been created with the said purpose in mind. At the very outset, the HGP made it clear that data obtained from HGP-funded research must be publicly available. Such efforts are based on the idea that our ability to expeditiously and effectively increase our knowledge of genetics depends on the ability of researchers to access current information. However, those who are responsible for creating and funding the HGP have a subsidiary goal of creating the technology for economic benefit.[74] As a result, the creation and upkeep of private genomic databases have begun. This has compelled the members of scientific community to think seriously about the impact of such databases on research ventures.

### 5.4.2 Accessibility of Abstract Genomic Data: Current Standard

A key issue relating to the data generated by the Human Genome Project was the promotion and encouragement of rapid sharing of the data.[75] The sharing of data was considered essential to foster the project, to avoid duplication and to expedite research in other areas.[76] At the other end, rapid data release conflicts with the fundamental scientific incentive to be the first to publish an analysis of one's data.[77] Since, biomedical research is increasingly data—rather than hypothesis—driven, a scientist could use his data collection more than once in the competition for publication and new grants. Here, it becomes pertinent to give the producers time to verify and validate their data and to gain some advantage from the effort they invested.[78]

---

[73] A. Marks and Karen K. Steinberg. (2002) The ethics of access to online genetic databases: private or public? Available at http://citeseerx.ist.psu.edu/viewdoc/summary?doi=10.1.1.102.4648 (last visited on 12 May 2012).

[74] *Ibid.*

[75] Bovenberg (2009).

[76] *Ibid.*

[77] *Ibid.*

[78] *Ibid.*

It was recognised that in order to move the results of genomic research from the bench to the marketplace, intellectual property protection for some of the data would be required. In this scenario, one of the leading scientists behind the international sequencing effort, Sir John Sulston, felt the need to get some kind of commitment from the international sequencing community that genomic information would be made publicly available.[79] He had noted in his personal account of the HGP that a gold rush mentality had evolved towards the genome, fostering a new breed of genome-based private companies.[80] Sulston's concern was reflected in the controversy surrounding the patenting by Myriad Genetics of the BRCA2-gene.[81]

#### 5.4.2.1 Bermuda Principle

Besides the pressure from the commercial sector, there was also a concern that large-scale sequencing centres, funded for the public good, would establish a privileged position in the exploitation and control of human sequence information. To address this issue, Sulston and his American Counterpart Bob Waterson organised a meeting in Bermuda, which was sponsored by the Wellcome Trust, a major supporter of the UK sequencing efforts.[82] At the Bermuda meeting, Sulston scribbled on a white board:

> Human Genomic Sequence generated at large-scale center:
> *RELEASE*
> Automatic release of sequence assemblies > 1 kb (preferably daily) Immediate submission of finished annotated sequence Aim to have all sequence freely available and in public domain for both research and development, in order to maximise its benefits to society
> *POLICY*
> The funding agencies are urged to foster these policies.[83]

Sulston was surprised to see that all attendants agreed to what be known as the Bermuda Principles.[84]

---

[79] *Ibid.*

[80] *Ibid.*

[81] *Ibid.*

[82] *Id.* at 339; (The attendants included sequencers—including Craig Ventor, who by then had not yet started his privately financed venture to sequence the human genome—and representatives from the WELLCOME Trust, the UK Medical Research Council, the NIH, NCHGR (National Center for Human Genome Research), DOE (US Department of Energy), the German Human Genome Programme, the European Commission, HUGO (Human Genome Organisation) and the Human Genome Project of Japan.).

[83] *Ibid.*

[84] *Ibid.*

## 5.4.2.2 Extension to Community Resource Projects

Since their adoption, the Bermuda Principles have been used as a point of reference for publicly funded large scale sequencing projects.[85] At a meeting in 2003, convened by the Wellcome Trust, an international group of data producers, users, database personnel, journal-editors and funding agency representatives unanimously agreed that pre-publication release of large-scale genome sequence data has been of tremendous benefit to the scientific research community. The group reaffirmed the Bermuda Principles and recommended that they be extended to all types of sequence data.[86]

Furthermore, the attendants also recognised that the data and other resources, which would be generated by 'community resource projects', should also be rapidly released to the community in an unrestricted manner. A 'community resource project' was defined as 'a research project specifically devised and implemented to create a set of data, reagents or other materials whose primary utility will be as a resource for the broad scientific community.'[87] Some community resource projects are International Human Genome sequencing Consortium, the Single nucleotide polymers (SNP) consortium, the International HapMap Project and, recently, the Encyclopaedia of DNA elements (ENCODE) Project carried out by the National Genome Research Institute.[88]

The attendants observed that these projects have become increasingly important as 'drivers of progress in biomedical research' and that immediate pre-publication release of the project data would serve the scientific community best.[89] They, however, realised that the pre-publication release model could jeopardise the standard scientific practice that the producers of the primary data should have both the right and responsibility to publish the work in a peer reviewed journal. Under a policy of rapid, unrestricted data release, there is a great risk that second comers who have not produced any data can grab the data sets posted on the web, analyse them and publish their results. As a result, the primary producers of the data can no longer publish these analyses themselves.[90] The introduction of publication exemption has been seen as viable solution. With the publication exemption, the scientific community would have the permission to use the unpublished data for all purposes, except the purpose of publishing the results of a complete genome sequence assembly or other large-scale analyses prior to the sequence producer's initial publication.[91]

The attendants concluded that the risk that sequence data were occasionally used in ways that violate normal standards of scientific etiquette was a necessary risk, to

---

[85] *Ibid.*
[86] *Id.* at 339–340.
[87] *Id.* at 340.
[88] *Ibid.*
[89] *Ibid.*
[90] *Ibid.*
[91] *Id.* at 340–341.

be accepted in view of the considerable benefit of immediate release. Instead of a publication exception, they encouraged producers to recognise this risk.[92]

### 5.4.2.3 Extension of Bermuda Principles to Phenotype Data

The attendants of the Wellcome Trust meeting noticed that apart from large scale 'community resource projects' many valuable small-scale data sets could come from other sources. Unlike large scale projects, those resources emerge from research efforts the primary goal of which is not resource generation, therefore, the contribution of their data to the public domain is more a voluntary matter. However, a clear benefit can be derived if these 'small-scale' data is converted into community resources as rapidly as possible.[93] Furthermore, the producers of such data should release them into the public domain voluntarily.[94]

### 5.4.2.4 Accessibility of Abstract (Post)Genomic Data: Developments

Although, Bermuda Principle deals with the problem of free riding by other scientists, however, it fails to cope up with the problem of free riding by commercial researchers. The International HapMap Project also suffers from problem of free riding by commercial researcher. The project is a multi-country effort to identify and catalogue genetic similarities and differences in human being. Although the project satisfies the definition of a community resource projects (CRP), however, the way in which it was to generate its data made it vulnerable for 'parasitic patenting'.[95] Genotype data produced during the early stages of the project would make it possible for other parties to construct haplotypes by combining the project's data with their own. As a result, these other parties could then file for patents on those derived haplotypes, and in doing so potentially restrict others from using those haplotypes and underlying data.[96] To remedy this situation, the HapMap Project developed a licensing strategy in the form of the HapMap Click Wrap License. Under this licensing agreement, scientists must indicate acceptance of the HapMap Click Wrap License Agreement in order to register for access to the said project.[97] Scientists can get a non-exclusive license to access by clicking on the acceptance button and conduct queries of the Genotype Database. The license allows researchers and scientists to copy, extract, distribute or otherwise use copies of the whole or any part

---

[92] *Id.* at 341.
[93] *Id.* at 342.
[94] *Ibid.*
[95] *Id.* at 347.
[96] *Ibid.*
[97] *Id*, at 347–348.

of the Genotype Database's data, in any medium and for all purposes, including commercial purposes.[98]

The rationale behind the Click Wrap License suggests that the need for the license requirement will disappear once the HapMap Project has ensured that most of the data it has generated have been placed in the public domain. This task was completed on 10 December 2004, when the project released all data publicly. As a result, the Project abandoned the click wrap requirement for access.[99]

#### 5.4.2.5 The Limitations of a License

Though click-wrap license seems an efficacious remedy to ensure unrestricted access to the data produced by Community Resource Projects, it is a license which only operates between the parties to the licensing agreement i.e. the licensor (the CRP consortium) and the licensee.[100] Secondly, even between the licensor and the licensee, it is unclear whether the condition of the license could be effectively enforced in case of a violation of the terms of the license. On the practical level, if the licensee breaches the agreement by filing a patent application, he will face an action for breach of contract. However, it is unlikely that such a breach would also invalidate the patent application or the patent issued. Here, the situation demands a more robust positive right to protect access to the database they produce.[101]

## 5.5 Open Source Analogy to Biotechnology

Open development movement is a recent phenomenon in the field of biotechnology. It is seen as a viable mean to ensure access to genetic information for researchers and scientists. It is also known as Open Bio movement. This movement has been inspired by the success of free and open source software (FOSS) movement.[102] There are some parallels between FOSS movement and Open-Bio movement. Both the movements are 'reactions to the proliferation of IPRs and to concerns that IPRs may restrict research and access to new innovations.'[103] Observers in both the fields have claimed that IPRs are quite often granted for inappropriate subject matter: preexisting art, or pure science, or even pure mathematics. For example, in software, certain innovations appear to be essentially mathematical algorithms, and in biotech, certain innovations appear to be essentially scientific discoveries.[104] In both

---

[98] *Id.* at 348.
[99] *Ibid.*
[100] *Id.* at 348–349.
[101] *Id.* at 350.
[102] Issac and Park (2009).
[103] *Ibid.*
[104] *Ibid.*

the fields, it is sometimes difficult to make the distinction between basic and applied research and development (R&D).[105]

### 5.5.1 Nature and Scope of Open Biotechnology

Open biotechnology has been used to refer to such different projects as an open journal (e.g. Public Library of Science), a new bioinformatics tool (e.g. the BioMoby messaging standard), a database (e.g. NIH db GaP), a big science project (e.g. HapMap or the Human Genome Project), a project to facilitate access to biotech research tools (Cambia BiOS) or a combination of these.[106] Taking advantage of the movement's popularity, many projects have been termed as open biotechnology.[107] Yann Joly proposes that, at a minimum, an open biotechnology project should meet the following criteria:

1. Make use at one stage or another of the internet and other information technologies (e.g. to promote quicker dissemination of results, promote collaboration and/or to improve project coordination).
2. Be designed in a way that will permit other members of the scientific community to collaborate on the project.
3. Include a strategy to ensure rapid public dissemination of the information and research results it generates.
4. Permit members of the scientific community to use its results without having to conclude restrictive agreements that would limit research freedom and integrity.
5. Not use intellectual property (IP) to limit access to the project, its results or to discriminate between different uses or different users.[108]

Joly emphasised on the possibility that an open biotechnology could also include a mechanism to allow the initial researchers to recuperate reasonable production costs invested in its realisation. He, however, cautioned that such mechanism should not impede the open nature of the project.[109] In the light of the above broad criteria, it can be inferred that open biotechnology is not necessarily antagonistic to IP and it is to develop an open source project that would make use of the patent system. He suggests that a variety of licensing schemes with or without IP (e.g. patent pool, non-assertion covenants, public domain, protected commons agreement, contractual licenses) can theoretically be used as the engine to support the open nature of the project.[110]

---

[105] *Ibid.*
[106] Joly (2010).
[107] *Ibid.*
[108] *Ibid.*
[109] *Ibid.*
[110] *Ibid.*

## 5.5 Open Source Analogy to Biotechnology

*Difference Between Information Technology and Biotechnology as to the IP Protection* Unlike IT, where most of the software is protectable through copyright; products of biotech are usually protected through the patent system. However, biotech inventions, which were protected through the patent system, have been found not deserving of such reward in recent legal decisions, forcing developers to rely on other weaker IP rights (e.g. copyrights, sui generis database rights), contractual law or commercial secrecy for protection.[111] As a result, it becomes difficult to develop simple licensing models to ensure the openness of a given project.[112]

In the context of patentable goods, the question arises, can the patent system be used, as copyright is, to ensure open development and access? The high cost and legal uncertainty involved in genetic patents create doubt as to the viability of such an approach.[113] It can be inferred from this that any inventor who relies on an open patent license would need to charge a sufficient cost to its licensees to recuperate its investment in the patent.[114] This would act as deterrence for potential users seeking a license.[115] This goes against the small projects, which cannot afford the cost of patents. They prefer to rely on commercial secrecy to protect their inventions. Some suggests umbrella organisation as a viable solution that could assume the responsibility for maintaining and protecting donated patents for researchers.[116] Since patentability of an increasing number of basic research findings has become cost intensive and legally uncertain, a fair number of scientists have chosen contractual licenses as a viable mean to ensure open or controlled access to research results to scientific community. Although these licenses seem less expensive and easier to design as compared to patent licenses, however, they do not seem effective against third parties to the original contract.[117]

There has been a growing tendency among the scientists to extend the protection of contractual licenses to those goods that are not protectable through IP (e.g. natural phenomena or raw data). Such use of contractual licenses may have a paradoxical effect of limiting access to an already public good to protect open access. Some suggest a strategy of leaving the goods in the public domain; however, the problem with this approach is that the goods remain vulnerable to abuse from commercially minded parties. In such a situation, 'large biopharmaceutical companies could access the good, modify it in small ways and use IP to control and market it, restricting its future use by members of the scientific community.'[118] Most open biotechnology projects have to deal with the sheer complexity of existing licenses and access agreements.[119]

---

[111] *Id.* at 418; See *In re: Dane K. Fisher and Raghunath v. Lalgudi* 421 F.3d 1365 (Fed. Cir. 2005), *In re: Marek Z. Kubin and Raymond G. Goodwin* No. 09–667, 859 (Fed. Cir. April 3, 2009).
[112] *Ibid.*
[113] *Ibid.*
[114] *Ibid.*
[115] *Ibid.*
[116] *Ibid.*
[117] *Ibid.*
[118] *Ibid.*
[119] *Ibid.*

Generally, small bioinformatics software projects operate under an open source model. Participants consider this model as an efficient mechanism for information dissemination, reduction of duplicative effort, and rapid development of software. However, they do not believe that all bioinformatics software should be open source.[120]

Open biotechnology movement is still in its infancy; however, it is gaining support from policymakers, NGOs and research funders.[121] Simple, efficient and legally valid open licenses are necessary for the success of Open bio movement.[122] Yann Joly suggests that in order to 'streamline and standardize current efforts, the creation of an international association where researchers interested in open biotechnology licensing could discuss common problems and harmonize their efforts would be very beneficial.'[123]

### 5.5.2 *Difference Between Open Source Software and Open Source Bioinformatics Software*

Most of the bioinformatics softwares are publicly funded while open source software is private. Most of the research universities prescribe a necessary condition that employee rights in software developed by using university resources are assigned to the university. Here, the policy of the universities towards open source software development becomes quite relevant.[124] For instance, in the USA, the policies of University of Washington and Georgia State regarding open source software depend upon whether it is perceived as commercially valuable. In the case of software, which is not commercially valuable, the research preference governs. On the other hand, in the case of commercially valuable software, both universities recommend that software and source code be licensed free of charge to non-commercial users. Since the differentiation between commercial and non-commercial can be maintained only through limits on redistribution of source code, such limits are in tension with open source principles that counsel against such limits.[125]

### 5.5.3 *Genomic Database Projects: The Human Genome Project as Open and Collaborative Genomic Database*

The Human Genome Project was probably the first most important open and collaborative genomic database.[126] The producers of the human genome sequence did not

---

[120] Rai (2005).
[121] Joly, *supra* note 107.
[122] *Ibid.*
[123] *Ibid.*
[124] Rai, *supra* note 121. at 141.
[125] *Ibid.*
[126] *Ibid.*

## 5.5 Open Source Analogy to Biotechnology

simply put the raw data into the public domain. However, an open source software programme, known as the distributed annotation system (DAS), was set up to facilitate collaborative improvement and annotation of the genome.[127] The DAS system allows any interested party to set up an annotation server. It enables end users of the information to choose the annotations they want to view by typing in the URLs of the appropriate servers.[128] One of the designers of the DAS system, Lincoln Stein, explained that DAS was designed to facilitate comparisons of annotations among several groups. The ultimate purpose behind the DAS system is that an annotation that is similar among multiple groups will be more reliable than an annotation that is noted by one group.[129]

The policy relating to the data dissemination and improvement policy of the HGP and other large scale genome mapping projects was generally developed by the National Institute of Health (NIH) and essentially imposed on the administrators of the participating universities. Although universities do not play a significant role in formulating the policy, they seem to have acquiesced in the rejection of proprietary rights.[130] This was the reason why the NIH did not invoke the cumbersome legal procedure set up by Bayh-Dole to restrain university patenting.[131]

In order to check the private effort of Craig Venter to sequence the genome, there were discussions within the HGP over using some type of 'copyleft'[132] license on the data produced by the project. Copyleft license was suggested with the possibility that it would prevent private entities, particularly Craig Venter from gaining advantage over the public data, by making proprietary any improvements Celera made to the public data.[133] Although the HGP leaders rejected a copyleft approach, however, National Human Genome Research Institute (NHGRI) along with other funding organisations embraced a copyleft style policy in setting up the International Haplotype Mapping Project (HapMap). The underlying object of this project is to catalogue haplotypes (patterns of genetic variation) and link such patterns to disease phenotypes.[134] Arti K. Rai suggests that 'at a minimum, open source software may be a good alternative for producing an output of reasonable quality at low cost.'[135]

---

[127] *Id.* at 142.

[128] *Ibid.*

[129] *Ibid.*

[130] *Ibid.*

[131] *Ibid.*

[132] The right to freely use, modify, copy and share software works of art etc. on the condition that these rights be granted to all subsequent users and owners of copyleft. Dictionary.com. The Free On-line Dictionary of Computing. Denis Howe, available at http://dictionary.reference.com/browse/copyleft (visited on 28 June 2012).

[133] Rai, *supra* note 121.at 147.

[134] *Ibid.*

[135] *Id.* at 145.

### 5.5.4 Importance of Open and Collaborative Databases

Open and collaborative approach to database generation is of immense value as it not only allows comprehensive database annotation, but it also provides an 'infrastructure' of freely available scientific information that all researchers, including wet-lab researchers, can use.[136] As compared to software projects database generation can require substantial capital investment.[137] Despite the fact that data generation can be open and collaborative, some restriction on participation, as well as public funding, will generally be necessary.[138] The high cost required for generating initial data, and the corresponding value associated with such data, public funding of databases probably undermines the ability of private businesses to form around databases. Unlike software, it is unlikely that private database businesses can be built on a service model.[139]

### 5.5.5 Open Standards

Several biotechnology standards are focussing on various kinds of data exchanges viz. the Clinical Data Interchange Standards Consortium (CDISC), which develops industry standards to promote medical and biopharmaceutical product development; the Open Bioinformatics Foundation (OBF), a volunteer organisation, which focuses on supported open source programming projects in bioinformatics.[140] Popular supported projects are BioMOBY and BioDas. BioMOBY denotes a small, grant-funded open source project dedicated to the creation of standards and tools for the registration and exchange of biological data stored on multiple hosts. The BioDas project supports the development of an open source distributed annotation system (DAS) for exchanging and collecting annotations on genomic sequence data.[141] Technology standards ensuring transparent and reasonable licensing terms can reduce patent thickets problem.[142]

Industry groups in standards development organisations (SDOs) will have to reconcile the interest of intellectual property owners with the interests of others who wish to practice the standard. It can be inferred from this that SDOs should make it necessary for the participants to agree and license all patents essential for compliance with the standard on 'fair, reasonable and non-discriminatory' terms. In the absence of such safeguard, there is always a risk that SDO's standards will be captured by a strategic member.[143]

---

[136] *Id.* at 147.
[137] *Ibid.*
[138] *Ibid.*
[139] *Ibid.*
[140] Issac and Park, *supra* note 103 at 235.
[141] *Id.* at 235–236.
[142] *Id.* at 236.
[143] *Ibid.*

5.5 Open Source Analogy to Biotechnology 189

Free and open standards are seen as a viable mean to avoid such risk of capture. Free and open standard describes a situation, where any party is licensed to read and implement it without payment. Free and open standards can benefit but the benefits appear especially great for new entrants with small or non-existent IP portfolios.[144]

The most preeminent open standards body of the internet is the World Wide Web (W3C). W3C standards are focussed on data exchange and display. Adoption of W3C standards has been extremely widespread, ensuring a remarkable level of interoperability on the internet.[145] There is no such a single SDO that plays a comparable role to the W3C in biotechnology.[146] However, efforts are being made to promote open standards, especially in the areas of data exchange and interoperability.[147]

### 5.5.6 Free and Open Development

Free and open development describes a situation, where innovations are shared freely rather than fenced off with IP claims. Distribution of a modified technology may be regulated in such a way which ensures that the modification also remains free and open.[148] Here, enabling disclosure is also public. Although IP protection may be sought and granted, however, the licensing to use, redistribute and modify the technology is provided *gratis* by the developer. This is why while an enabled technology in the public domain is obviously free and open, patented and copyrighted technology may be as well.[149] The most popular example of free and open development is the FOSS development. A software development is treated to be free and open only if the source code is readily available and freely redistributable. In such software, there are no legal restrictions on the redistribution of the unmodified open source software to others. Practically, the FOSS has generally been available for download without charge.[150]

Free and open development practices have generally been modelled on the FOSS example. This is remarkably apparent in the field of bioinformatics, where many FO development efforts are FOSS development efforts.[151] The well-known FOSS software tools from the Linux operating system to Python scripting language often play a supportive role in bioinformatics.[152]

---

[144] *Ibid.*

[145] *Id.* at 136–137.

[146] *Id.* at 237.

[147] *Ibid.* (For instance, the standards developed by CDISC, which reflect its mission of improved data quality and accelerated innovation in medicine and biopharmaceuticals. CDISC explicitly commits to vender-neutral, platform independent standards.).

[148] *Ibid.*

[149] *Ibid.*

[150] *Id.* at 238.

[151] *Ibid.*

[152] *Ibid.*

Databases have also been induced by the open development paradigms. The best example of the open and collaborative database project is the International HapMap project. The project focuses upon the mapping of common patterns of variations in the genome. The ultimate object of the project is to put the completed data in the public domain. The data access policy of the project maintains that users agree not to reduce other's access to the data and to share the data only with others who have made the same agreement. Patenting of subsequent discoveries is allowed 'as long as patentees do not prevent others from obtaining access to the project's data.'[153] The SNP Consortium is a perfect example of open development, which provides tools to advance industry goals.[154] Under the SNP Consortium, several large pharmaceutical and technology companies have joined hands with Wellcome Trust and academic researchers to file patent applications on SNPs that will be freely accessible to all.[155]

### 5.5.7 Whether Open Development Is a Viable Business Model

In the business field, a producer may find that open development in one area can stimulate product demand and minimise costs in other area. Open development, therefore, may enhance the rate of innovation in product complementary to a firm's commercial enterprises. For example, open may pay freely to reveal innovations in razors if this stimulates additional razor innovations that increase the demand for blades. Or a firm may freely reveal an innovation in order to stimulate the development of a research tool it needs in a separate commercial endeavour.[156] In an open development environment, a collective invention process ensures the free exchange of information among researchers about technology.[157] Furthermore, users of research tools often recommend improvements in functionality or interface. They may trust on open revelation to encourage the adoption of innovations that might otherwise be ignored.[158]

Open Bio has certain merits as well as demerits. It promotes access to certain research tools and encourages user innovation that might be absent from a different development process, however, it may also reduce the profits from commercial R&D in the area of research tools.[159] The overall impact on innovation depends on conflicting influences: 'an open innovation process may lower the cost of research tool innovation by eliminating the transaction costs of license negotiations, but the

---

[153] *Ibid.*
[154] *Ibid.*
[155] *Ibid.*
[156] *Id.* at 239.
[157] *Ibid.*
[158] *Id.* at 240.
[159] *Ibid.*

potential benefits of inventive activity may no longer include possible profits from licensing the innovation.'[160]

One of the public policy justifications for patent is that it stimulates innovation and diffusion by raising the private return to research, development and commercialisation. If open development lowers this private return, growth may suffer.[161]

### 5.5.8 Is the Open Source Analogy Relevant to Biotechnology?

Open source analogy has been applied to biotechnology in order to harness the communication, licensing and organisational innovation. For example, BIOS Initiative attempts to 'extend the metaphor and concepts of open source software' to biotech innovation.[162]

**Analogies Between Biotech and Software** In the case of Free and open source software, enterprise quality software is being produced as a hobby by amateurs, perhaps even by teenage hackers. On the contrary, some biotechnology fields require a team of scientists with advanced degrees, and the credentials of scientists and engineers matter.[163]

Some biotech research resembles with software development e.g. some research in computational biology focuses on algorithm development, which is in essence, a software development. Other biotech related research emphasises on facilitating data exchange, which bears analogy to the W3C efforts to develop standards for information exchange on the web. Experiences gained from the FOSS movement are most likely to be applicable to such neighbours. Apart from similarities, there are significant differences.[164]

**Disanalogies Between Biotechnology and Software** Open development is generally not a matter of source code sharing outside the realm of bioinformatics.[165] In the field of biotechnology, once we move outside from the bioinformatics realm, 'open source' becomes largely a metaphor for open development: sharing the underlying technological secrets or information and giving access.[166]

---

[160] *Ibid.*

[161] *Id.* at 241. (In a simple endogenous growth model, Saint-Paul shows that philanthropic innovation can reduce long run growth by crowding out proprietary innovation.).

[162] *Ibid.*

[163] *Ibid.*

[164] *Id.* at 242.

[165] *Ibid.*

[166] *Id.* at 242–243.; In particular, the 'open source' metaphor is misleading if it is applied to the 'code' in biomedical sequences, including ESTs, SNPs or even genes. If these are patented, the code in such cases goes far beyond the standard applied in software patents, where software patents often have to reveal the underlying source code in order to satisfy the enablement requirement.

## 5.5.9 Innovation and Open Development

Biomedical research is increasingly proprietary and secretive, creating fears that future progress may be impeded by restricted access and licensing difficulties. Some recommend easy access to certain type of data and research tools to researchers, while others propose open development. 'Open and Collaborative' science provides relief from potential patent thickets or problems of hold up. However, there is a possibility that open development may reduce innovation effort by reducing anticipated profits.[167] At present, any assessment of these possibilities would be extremely tentative.[168] It can be said conclusively that at the moment open development is in practice and is working in a number of areas, including operating systems, scripting languages and sequencing algorithms. Bioinformatics databases also illustrate additional possibilities for free and open development.[169] Prospects for Open Bio look much better as we get closer to basic R&D, projects involving platforms or enabling technologies.[170]

## 5.6 Is Open Bio Good for Developing Countries?

Open Bio have different impacts on developed and developing countries. Open Bio may provide developing countries the opportunity to imitate, learn and innovate without violating their IP agreements. Here, Open Bio effectively lowers the cost of entry into biotech research.[171] Maurer, Rai and Sali propose that open source software can provide a model for improving innovation in tropical medicine. Since 'open source' discoveries will not be patented, it is hoped that zero licensing fees and competitive pressures will conspire to keep prices low. Clearly, the choice of license will be crucial.[172]

India made an effort to pre-empt biopiracy by digitally placing traditional knowledge into the public domain. Central to this effort is the Traditional Knowledge Digital Library (TKDL), under the auspices of India's National Institute of Science, Communication and information Resources (NISCAIR).[173]

Issac and Park believe that Open Bio may increase the access to data and research tools for developing countries.[174] Since Open Bio developments are freely avail-

---

[167] *Ibid.*

[168] *Id.* at 244.

[169] *Ibid.* (From an economist's perspective, one of the most fascinating developments in Open Bio is the SNP Consortium, where commercial interests were driven to open development in an effort to reduce future transaction costs.).

[170] *Ibid.*

[171] *Id.* at 245.

[172] *Ibid.*; See Stephen M. Maurer, Arti Rai et al. Finding cures for tropical diseases: Is open source an answer? PLos Medicine 1: 183–186.

[173] *Ibid.*

[174] *Id.* at 248.

able to developing countries, therefore it reduces the pressure on those countries to transgress recently harmonised IPR standards. Issac and Park are sceptical that on its own, Open Bio will lend much stimulus to the development of the biopharmaceutical innovations so desperately needed by the developing world. However, in the present age, they expect Open Bio to work best when complementary to targeted government research support. A detailed exploration of this complementarity is an important area for future research.[175]

# References

Bovenberg Jasper A (2009) Accessibility of biological data: a role for the European database right. In: Castle David (ed), The role of intellectual property rights in biotechnology innovation. Edward Elgar Publishing Limited, Massachusetts, U.S.A., p. 338

Gopalan Raghuvaran (2009) Bioinformatics: scope of intellectual property protection. Journal of Intellectual Property Rights 14: 46

Hultquist Steven J, Harrison Robert and Yang Yongzhi (2002) Patenting bioinformatics inventions: emerging trends in Europe. Nature Biotechnology 20: 517

Hultquist Steven J, Harrison Robert and Yang Yongzhi (2003) Patenting bioinformatics inventions: emerging trends in the United States and Europe. IPTL notes 5. Available at http://www.iptl.com/articles/iptlnotes_winter03.pdf. Accessed 20 April 2011

Issac Alan G and Park Walter G (2009) Open development: is the 'open source' analogy relevant to biotechnology? In: Castle David (ed), The role of intellectual property rights in biotechnology innovation. Edward Elgar Publishing Limited, Massachusetts, U.S.A., p. 226

Joly Yann (2010) Open biotechnology: licenses needed. Nature Biotechnology 28: 417

Marks A and Steinberg Karen K (2002) The ethics of access to online genetic databases: private or public? Available at http://citeseerx.ist.psu.edu/viewdoc/summary?doi=10.1.1.102.4648. Accessed 12 May 2012

Maurer Stephen M, Rai Arti and Andrej Sali (2004) Finding cures for tropical diseases: is open source an answer? Plos Medicine 1: 183–186

Rai Arti K (2005) Open and collaborative research: a new model for biomedicine. Intellectual Property Rights in Frontier Industries 140. http://scholarship.law.duke.edu/faculty_scholarship/882. Accessed 12 May 2012

Rees Dianne (2003) Patenting bioinformatics tools. www.bio-itworld.com. Accessed 10 May 2012

---

[175] *Id.* at 248–249.

# Chapter 6
# Implications of Genetic Patents on Human Genetic Resources: Issues of Ownership, Benefit Sharing and Informed Consent

With the increasing commercialisation of human genetic research, human genetic material has become a source for patenting. The increased extension of patent rights to human genetic material has serious implications for research subjects and patients whose genetic material is used in the research. The ownership of human genetic material has become a controversial issue as a variety of proprietary rights are claimed over it such as patent rights, personal rights, sovereign rights and academic rights. Among these claims, the ownership rights claims of patients, researchers and research subjects has been contested in courts and started an open debate among legal scholars and policy makers. In genetic research, researchers and sponsors thereof are basically concerned to exploit the human genetic material to earn credit by claiming their exclusive rights. It raises the question, whether may researchers obtain patent rights through observation, isolation and manipulation of the human genetic material, without recognising and admitting contribution of research subjects and patients who have given their genetic material for the research? The commercial exploitation of human genetic material is not confined to an individual but it also extends to human genetic resources of countries. TRIPS does not contain any explicit reference to genetic material, and the laws that restrict access to genetic material to obtain remuneration for the nation such as Convention on Biological Diversity (CBD) exclude human genetic material from their ambit. This has led to a growing exploitation of human genetic resources for scientific or commercial purposes.

## 6.1 Ownership of Human Genetic Material

There exists great confusion regarding the ownership of extracted human genetic material as a variety of proprietary rights is claimed over it. A person can own his genetic material when it remains inside the human body; however, once the material is extracted from the body, the confusion as to the ownership rights begin. Initially, human genetic material was considered as product of nature and the issue of ownership relating to it had not created any significant problem. The issue, however, has become complicated since 1953, after the discovery of DNA's molecular structure by Watson and Crick. Due to rapid developments in the field of molecular biology and genetics, human DNA has evolved from the 'Secret of Life to a potentially valuable Commodity'.[1] Modern biotech techniques such as recombinant technology have made it possible to transform human genetic materials from products of nature to products of man. A human DNA now can be isolated, amplified, recombined and converted into valuable products such as human insulin or protein.[2]

The potential commodification of human genetic material occurs in different ways such as, it can be collected, stored and used for diagnostic and therapeutic purposes; it can also be sequenced and stored in databases, where it can be linked with associated health, clinical, environmental and life-style data to promote useful studies regarding complex diseases.[3] These various ways of commodification lead to conflicting perspectives as to the ownership of human genetic material. A variety of proprietary claims are being made on human genetic material. Human DNA is claimed as universal property shared by the entire human race. Intellectual property claims are also being made with respect to human gene as patenting of isolated and purified human gene has become a norm. In addition to patents, various other forms of intellectual property rights such as copyright and sui generis database rights are being claimed on human genome including human genome sequence data.[4] The rapid growth of DNA banks and gene banks encouraged various countries[5] to establish national health sector database. Further, in response to the aggressive patenting of human genetic resources, developing countries have sought to extend its sovereign control over human genetic resources. Here, human genetic resource is considered as a national property.[6] Moreover, academic researchers claim the blood genes and data they have collected in the course of their research as their academic property.[7] Due to the commodification of human genome in different ways, patients and research subjects are claiming personal proprietary claims on their genetic material used in research.[8] Individual patients stand for their personal property right claims on their blood gene and data. In addition to this, academic researchers claim blood, gene and data as their academic property, which they have collected in the course of research.

---

[1] Bovenberg (2006).

[2] *Ibid.*

[3] *Ibid.*

[4] *Id.* at 3.

[5] *Id.* at 3–4; (In 1998, the Parliament of Iceland passed the Act on the Centralized Health Sector Database. This Icelandic initiative has been followed by Estonian Gene Bank and the U.K. Gene Bank. The common object of these projects is to compile national or large scale collections of human biological material and to link these with associated clinical and personal health data and genealogical information).

[6] *Id.* at 4.

[7] *Id.* at 1.

[8] *Id.* at 4.

**Objections Against the Commercialisation of Human Genetic Material** Opponents to the commercialisation of human genetic material argue that human genome should not be controlled by market as it controls other external things. Another argument against the said commercialisation is that the God prepared the layout of the genetic sequence of human being and any modification in it should be considered as interference in the divine pattern.[9] The environmental concern is that 'you can't own a gene; to do so is to embrace a system in which nature, even our own nature, is to be manipulated, traded and commodified'.[10] Since technologically advanced countries are well placed to exploit the genetic resources of less developed and developing countries, which lack advance technologies, some see the extension of property rights over genetic resources by technologically advanced countries as a form of second colonial expansion.[11] The opponents consider private enclosure of human genetic material dreadful as it extends the market to traditionally foreign subject matters.[12]

**Human Genome: A Common Heritage to the Mankind** The objection against the commercialisation of human genome on the basis that it is the common heritage to the mankind shared by all had led to the adoption of United Nations Educational, Scientific and Cultural Organization's (UNESCO) *Universal Declaration on the Human Genome and Human Rights, 1997*. The Declaration was endorsed by the United Nations General Assembly on 9 December 1998, which proclaims that: 'The human genome underlies the fundamental unity of all members of the human family, as well as the recognition of their inherent dignity and diversity. In a symbolic sense, it is the heritage of humanity'.[13] This proclamation is considered as the cornerstone of the Declaration as all the subsequent provisions revolve around it. Another important provision regarding the commercialisation of human genome is Article 4, which provides that 'the human genome, in its natural state, shall not give rise to financial gains'.[14]

Here, the question arises as how and where to draw the line between natural and artificial forms of human genome. In common parlance, the words 'in its natural state' seem to refer to the human genome as it occurs in natural form, within the human body. However, it would be a limited interpretation of the terms which may restrict the scope of the Declaration. Under such interpretation even the raw genomic data that were being assembled by the Human Genome Project would not qualify as a common heritage subject matter. The Declaration does not give any clue as to how it intended to delineate between the natural and artificial.[15] However, the

---

[9] Boyle (2003).

[10] *Id.* at 102.

[11] *Id.* at 103.

[12] *Id.* at 98–99.

[13] Art. 1 of UNESCO's *Universal Declaration on Human Genome and Human Rights 1997*.

[14] Id.; Art. 2 of UNESCO's Universal Declaration on Human Genome and Human Rights 1997.

[15] Bovenberg, *supra* note 1 at 44.

landmark decision of the US Supreme Court in *Diamond versus Chakrabarty*[16] has restricted the product of nature doctrine by allowing patents for 'anything under the sun made by man'. After this decision, it has been established in various jurisdictions that isolated and purified DNA sequences can be the subject matter of patent applications.[17] The Declaration on Human Genome has incorporated the essence of *Diamond versus Chakrabarty* which seems to narrow down the scope of the human genome as common heritage subject matter to those sequences that have not been isolated and purified.[18]

## 6.2 Ownership Rights of Research Subjects over Their Genetic Material Used in Genetic Research

Researchers and sponsors of genetic research are well placed to commercially exploit the research carried on human genetic material of research subjects. They are capable of using genetic materials, information and other specimens collected during a primary research study for future unspecified research purposes.[19] The first human cells reproduced prolifically in laboratory setting was of a poor black woman, Henrietta Lacks, who died of a particularly virulent form of cervical cancer in 1950. Using Mrs. Lacks' cancer cells, scientists have reproduced 50 million t of her cells over time. Mrs. Lacks' cell line contributed to significant advancements in the field of life sciences, including the polio vaccine, cancer treatments and in vitro fertilization.[20] Despite Mrs. Lacks' cells' contribution to the development of the life science industry, her family has not shared any benefit.[21] This reflects the vulnerable condition of research subjects whose biological materials were used in profitable research without tendering them any benefit. The focus of this chapter is on the conflicting claims of researchers, research subjects and patients, paying much heed to the implications of patenting of human genetic material on the interests of research subjects and patients.

Great uncertainty persists as to what ownership rights patients and research subjects have in their genetic material. A study by the Office of Technology Assessment (OTA) reviewed the available legal, ethical and scientific literature relating to this issue and concluded that there was no certain answer to the questions of ownership and control.[22] The OTA report maintains that biological materials themselves have little use or value, until and unless scientists manipulate and use the materials to cell

---

[16] 447 U.S. 303, 309 (1980).

[17] Bovenberg, *supra* note 1 at 49.

[18] *Ibid.*

[19] Barnes and Heffernan (2004).

[20] Feldman (2011).

[21] *Ibid.*

[22] Office of Technology Assessment, *New Developments in Biotechnology: Ownership of Human Tissues and Cells* (March 1987).

lines, products or useful data.²³ Here, the question arises whether research subjects and patients who are supplying their genetic material for research purpose are making any contribution in the research or not.

The current trends in property and intellectual property laws are approaching to a point where 'anyone can have property rights in your excised cells except you'.²⁴ If a researcher works with a sample of human blood or tissue, the researcher, or the lab, has a property right on those cells. Likewise, if the researcher isolates a protein or a segment of DNA from the sample, the researcher, or the lab, has the property rights on the tangible isolated elements. Further, the sample or tissue or the cell lines developed from the tissue are treated under the contract law according to the agreements related to the transfer of tangible property, commonly called 'material transfer agreements'.²⁵ In addition to property rights, a researcher may also obtain intellectual property rights through observation, isolation and manipulation of the sample. However, much ambiguity exists as to the rights of individuals, who contributed the sample in the first place.²⁶ There are certain court decisions from the USA, which throw light upon various contentions involved in clinical research. Court proceedings have highlighted the uncertainty of rights and responsibilities in the area of ownership, control, and future uses of data and tissue gathered during research.²⁷

### 6.2.1 Moore versus Regents of the University of California

*Moore versus Regents of the University of California*²⁸ is a landmark case regarding the ownership of excised human tissues. John Moore was suffering from hair-cell leukaemia. He visited UCLA Medical Center for his treatment where his physician Dr. David W. Golde confirmed the disease and recommended him to undergo surgery to remove his spleen.²⁹ Dr. Golde informed Moore that he has reason to fear for his life, and the proposed splenectomy operation was necessary to slow down the progress of his disease.³⁰ Relying upon Golde's representation Moore signed a written consent form authorising the splenectomy. Consequently, his spleen was removed. Afterwards, he was asked by his physician to make several visits to the UCLA Medical Center. He made such visits at Golde's direction and based upon representations 'that such visits were necessary and required for his health and wellbeing, and based upon the trust inherent in and by virtue of the physician-patient

---

[23] Barnes and Heffernan, *supra* note 19 at 3.
[24] Feldman, *supra* note 20, at 1378.
[25] *Ibid.*
[26] *Id.* at 1379.
[27] Barnes and Heffernan, *supra* note 19 at 4.
[28] *Moore v. Regents of the University of California*, 51 Cal. 3d 120 (1990).
[29] *Id.* at 126.
[30] *Id.* at 127.

relationship...'.[31] At each visit his physician Dr. Golde withdrew additional samples of blood, blood serum, bone marrow aspirate and sperm. Moore subsequently learned that defendants including Dr. Golde were conducting research on his cells, and planned to benefit financially by exploiting the cells and their exclusive access to the cells by virtue of Golde's ongoing physician-patient relationship.[32] Golde established a cell line from Moore's T-lymphocytes, and Regents applied for a patent on the cell line, listing Golde and Quan as inventors. A patent[33] was issued naming Golde and Quan as the inventors of the cell line and the Regents as the assignee of the patent.[34]

Aggrieved by the act of defendants, Moore sued them under various causes of action. At the California Supreme Court, the case primarily revolved around two potential theories of liability. The first involved a claim for the conversion of the plaintiff's genetic material by the action of the defendant. This claim pre-supposed that the defendant had a property interest in his genetic material in the first place—and that it was not 'abandoned' during the course of surgery. The second theory was based on an analogy of the law of informed consent in medical malpractice and claimed that the defendants breached their fiduciary duty when they did not disclose the reasons why they continued to call the plaintiff back for more and more tests, all of which increased their available supplies of this critical cell lines, without disclosing to him their ulterior purpose.[35]

The court found that Moore had no property rights to his discarded cells or any profits made from them. In order to establish conversion claim, the plaintiff must establish an actual interference with his ownership or right of possession. In a case, where plaintiff neither has title to the property alleged to have been converted, nor possession thereof, he cannot maintain an action for conversion.[36]

In Moore's case, the court held that since the plaintiff did not expect to retain possession of his cells, he must have retained an ownership interest in them to sue for their conversion There are, however, several reasons to doubt that he retained such interest. First, there is no precedent in support of plaintiff's claim. Second, California statutes drastically limit any continuing interest of a patient in excised cells by requiring that they be destroyed after use. Third, the subject matter of the patent (i.e. the patented cell line and the technology and products derived from it) cannot be Moore's property.[37] Lacking direct authority for imparting the law of conversion in the present context, Moore relied primarily on decisions addressing privacy rights.[38]

---

[31] *Ibid*

[32] *Ibid.*

[33] U.S. Patent No. 4,438,032 (Mar. 20, 1984).

[34] *Moore, supra* note 28 at 128.

[35] Epstein (2011).

[36] *Moore, supra* note 28 at 137.

[37] *Id*. at 138.

[38] *Ibid.*

The next consideration that makes Moore's claim of ownership problematic is California statutory law, which drastically limits a patient's control over excised cells. Pursuant to Health and Safety Code section 7054.4, '[n]otwithstanding any other provision of law, recognizable anatomical parts, human tissues, anatomical human remains, or infectious waste following conclusion of scientific use shall be disposed of by interment, incineration, or any other method determined by the state department [of health services] to protect the public health and safety'.[39] Finally, the subject matter of the Regents' patent, the patented cell line and the products derived from it cannot be Moore's property. This is because the patented cell line is both factually and legally distinct from the cells taken from Moore's body. Federal law permits the patenting of organisms that represent the product of 'human ingenuity', but not naturally occurring organisms.[40] Thus, Moore's allegations that he owns the cell line and the products derived from it are inconsistent with the patent, which constitutes an authoritative determination that the cell line is the product of invention.[41]

The court considered Moore's claim to own the biological materials as a request to extend the conversion theory and opined that such claim demands express consideration of the policies to be served, extending liability rather than blind deference to a complaint alleging as a legal conclusion the existence of a cause of action.[42]

The court pointed out three reasons as to why it was inappropriate to impose liability for conversion based upon the allegations of Moore's complaint. First, a fair balancing of the relevant policy considerations counselled against extending the tort. Second, problems in this area were better suited to legislative resolution. Third, the tort of conversion was not necessary to protect patients' rights. For these reasons, the court concluded that the use of excised human cells in medical research does not amount to a conversion.[43]

The court also explained that since conversion is a strict liability tort, it would impose liability on all those into whose hands the cells come, whether or not the particular defendant participated in, or knew of, the inadequate disclosures that violated the patient's right to make an informed decision. On the contrary, the fiduciary duty, the fiduciary-duty and informed-consent theories protect the patient directly, without punishing innocent parties or creating disincentives to the conduct of socially beneficial research.[44] The court also observed that conversion law in such cases will impede the research by restricting access to the necessary raw materials.[45]

On this count, the California Supreme Court ruled that Moore's consent was not obtained, and the doctors were in breach of their fiduciary duty. The court held that 'a physician who is seeking a patient's consent for a medical procedure must,

---

[39] *Id.* at 141.
[40] *Id.* at 142–143.
[41] *Id.* at 143.
[42] *Ibid.*
[43] *Id.*, at 143–144.
[44] *Id.* at 145.
[45] *Ibid.*

in order to satisfy his fiduciary duty and to obtain the patient's informed consent, disclose personal interests unrelated to the patient's health, whether research or economic, that may affect his medical judgment'.[46]

The court affirmed that Moore's allegations state a cause of action for invading a legally protected interest of his patient. The court recognised that 'a cause of action can lie under the informed consent doctrine as a breach of the fiduciary duty to disclose material facts, or the lack of informed consent in obtaining consent to conduct medical procedures'.[47]

Arabian J. raised some deep moral issues involved in Moore's case while writing a concurring opinion. However, he suggested that the question whether Moore's cell should be treated as property susceptible to conversion could be better decided by the legislature than the court.[48]

Justice Broussard concurred in part and dissented in part.

While concurring with the majority decision, Broussard expressed his dissent over the decision that the facts alleged in Moore's case do not state a cause of action for conversion. He maintained that:

> ...Although a patient may not retain any legal interest in a body part after its removal when he has properly consented to its removal and use for scientific purposes, it is clear under California law that before a body part is removed it is the patient, rather than his doctor or hospital, who possesses the right to determine the use to which the body part will be put after removal. If, as alleged in this case, plaintiff's doctor improperly interfered with plaintiff's right to control the use of a body part by wrongfully withholding material information from him before its removal, under traditional common law principles plaintiff may maintain a conversion action to recover the economic value of the right to control the use of his body part. Accordingly, I dissent from the majority opinion insofar as it rejects plaintiff's conversion cause of action.[49]

While dissenting with the majority opinion over the interpretation of the term property, Mosk J. maintained: '[o]wnership is not a single concrete entity but a bundle of rights and privileges as well as of obligations'.[50] He further added that '[s]ince property or title is a complex bundle of rights, duties, powers and immunities, the pruning away of some or a great many of these elements does not entirely destroy the title...'.[51] Regarding the statutory restriction of California law, Mosk J. observed:

> The same rule applies to Moore's interest in his own body tissue: even if we assume that section 7054.4 limited the use and disposition of his excised tissue in the manner claimed by the majority, Moore nevertheless retained valuable rights in that tissue. Above all, at the time of its excision he at least had the right to do with his own tissue whatever the

---

[46] *Id.* at 132–133.

[47] *Moore v. Regents of the University of California*—Case Brief Summary, available at http://www.lawnix.com/cases/moore-regents-california.html. (last visited on May 22, 2012).

[48] *Moore, supra* note 28 at 150 (Arabian J., concurring).

[49] *Id.* at 152 (Broussard J., dissenting).

[50] *Id.* at 166 (Mosk J., dissenting) quoting *Union Oil Co. v. State Bd. of Equal.* (1963) 60 Cal.2d 441, 447 [34 Cal.Rptr. 872, 386 P.2d 496].

[51] *Id.* at 167 (Mosk J., dissenting); quoting *People v. Walker* (1939) 33 Cal. App. 2d 18, 20 [90 P.2d 854].

defendants did with it: i.e., he could have contracted with researchers and pharmaceutical companies to develop and exploit the vast commercial potential of his tissue and its products.[52]

Emphasising upon the contributions made by patients and researchers, Mosk J. quoted an excerpt from *University of California, Los Angeles Law Review*:

> Recognizing a donor's property rights would prevent unjust enrichment by giving monetary rewards to the donor and researcher proportionate to the value of their respective contributions. Biotechnology depends upon the contributions of both patients and researchers. If not for the patient's contribution of cells with unique attributes, the medical value of the bioengineered cells would be negligible. But for the physician's contribution of knowledge and skill in developing the cell product, the commercial value of the patient's cells would also be negligible. Failing to compensate the patient unjustly enriches the researcher because only the researcher's contribution is recognized.[53]

The most challenging feature of the Moore judgement was the decision on whether or not Moore's cells and tissues could be considered as property, thereby allowing them to be converted. The court, however, restricted itself to the question of proprietary rights over the cell line developed by the doctors at the medical institute.[54]

### 6.2.2 Greenberg versus Miami Children's Hospital (Canavan Disease Case)

In the field of clinical genetic research, it is argued that the discovery of disease related genes is increasingly the result of a pragmatic partnership between the researcher and those afflicted with the condition.[55] Research participants not only contribute in studies but they are involved with 'broader aspects of the research, including identifying and obtaining samples from other affected individuals, and even with securing research funds'.[56] The vulnerability of the research participants has been exposed in *Greenberg versus Miami Children's Hospital Research Institute Inc*.[57] popularly known as Canavan disease case.

In the present, the US District Court of Florida allowed plaintiffs (a group of individuals who provided genetic materials for medical research into Canavan's disease) to proceed with their claim for unjust enrichment against the principal investigator of the study, Dr. Matalon, and the research institution, Miami Children's Hospital.[58] Both parties to this case were jointly involved to detect and find a cure

---

[52] *Id.* at 167.

[53] *Id.* at 176, quoting "Toward the Right of Commerciality: Recognizing Property Rights in the Commercial Value of Human Tissue," 34 *UCLA L. Rev.* 230 (1986).

[54] N. Narayanan, "Patenting of human genetic material v. bioethics: revisiting the case of John Moore v. Regents of the University of California." available at http://www.ncbi.nlm.nih.gov/pubmed/20432879 (last visited on may 21, 2012).

[55] World Health Organization (2005).

[56] *Ibid.*

[57] *Greenberg v. Miami Children's Hospital,* 2003 WL 21246347 (S.D. Fla. May 29, 2003); 264 F. Supp. 2d 1064 (S.D. Fla. 2003).

[58] Barnes and Heffernan, *supra* note 19 at 4.

for Canavan disease. Canavan disease is a rare genetic disorder that occurs more frequently in Ashkenazi Jewish families.[59]

In order to develop tests to determine carriers of the disease, Greenberg (plaintiff) approached Dr. Matalon (defendant), a research physician, to discover the gene sequence of Canavan disease. He convinced Canavan families worldwide to donate tissue, blood and financial support to Matalon's project. The joint efforts of Greenberg and Matalon resulted in the development of a Canavan registry containing epidemiological, medical and other information about the families who had donated material. Later on, Matalon became affiliated with Miami Children's Hospital Research Institute (the Institute) (defendant) and Miami Children's Hospital (MCH) (defendant), but continued his research work on Canavan disease with Greenberg and the participating families by accepting additional tissue and blood samples and financial support.[60] In 1993, Matalon got breakthrough in isolating the gene responsible for Canavan disease. In the very next year, Matalon filed a patent application for the genetic sequence he identified. In this context, he had never informed Greenberg and the families about his intentions to seek a patent on the Canavan research. In 1997, Matalon was listed as the inventor of the gene patent, and he and MCH began restricting any activity related to Canavan disease without compensation.[61]

The plaintiffs (Greenberg and the Canavan families) filed a six-count complaint against Matalon and MCH alleging lack of informed consent, breach of fiduciary duty, unjust enrichment, fraudulent concealment, conversion and misappropriation of trade secrets and sought damages along with patent royalties. They claimed that all tissue, blood and financial support provided to Matalon was done for the purpose of carrier detection, identifying mutations of the disease, and providing testing that would benefit the public at large. Greenberg requested a permanent injunction restraining Matalon and the institute from enforcing the Canavan gene patent rights.[62]

Rejecting the breach of informed consent claim, the court held that although medical researchers have a duty of informed consent in certain circumstances, the court would not extend that duty to require disclosures to subjects of a researcher's economic interests in the research.[63] The court distinguished *Moore versus Regents of the University of California* from the present case on the ground that the researcher in the former case provided care to Mr. Moore, while Dr. Matalon did not provide care to any of the plaintiffs in later case.[64] The decision appeared to be based on the logic that the plaintiffs were better characterized as 'donors' of the genetic materials than as objects of human experimentation.[65]

---

[59] *Greenberg v. Miami Children's Hospital Research Institute Inc*, available at http://indylaw.indiana.edu/instructors/orentlicher/healthlw/Greenberg.htm (visited on May 30, 2012).
[60] *Ibid.*
[61] *Ibid.*
[62] *Ibid.*
[63] Barnes and Heffernan, *supra* note 19 at 4.
[64] *Greenberg, supra* note 59.
[65] Barnes and Heffernan, *supra* note 19 at 4.

The court, however, recognised the 'unjust enrichment' claim by maintaining that the complaint alleged more than just a donor-donee relationship and had stated an appropriate legal claim: namely, that the plaintiffs would not have donated their genetic material if they had known of the researcher's goal of financial gain from the results of the research and that due to the subjects' ignorance of the researcher's intentions, the researcher and his hospital had unfairly gained a financial benefit.[66] The court, therefore allowed the plaintiffs to proceed with their claim for unjust enrichment against the defendants.[67]

Defendants moved to dismiss the breach of fiduciary duty count by contending that plaintiffs did not plead the elements of fiduciary duty and mentioned that Florida courts have found fiduciary relationships in this context 'when confidence is reposed by one party and a trust accepted by the other'.[68] They contended that the complaint does not allege any facts that show the trust was recognised and accepted. On the other hand, the plaintiffs alleged that defendants accepted the trust by undertaking research that they represented as being for the benefit of the plaintiffs.[69] After considering the whole facts, the court found that plaintiffs have not sufficiently alleged the second element of acceptance of trust by defendants and therefore have failed to state a claim. The court maintained that there is no automatic fiduciary relationship that attaches when a researcher accepts medical donations and the acceptance of trust, the second constitutive element of finding a fiduciary duty, cannot be assumed once a donation is given.[70]

While rejecting the fraudulent concealment claim of plaintiffs, the court made it clear that the absence of a fiduciary relationship precluded any duty of disclosure to the plaintiffs; that the defendants' patent was a matter of public record, therefore not fraudulently concealed; and that plaintiffs did not suffer any economic injury.[71]

In the conversion claim, plaintiffs alleged that they had a property interest in their body tissue and genetic information. Regarding the conversion allegation, they specifically mentioned that they owned the Canavan registry in Illinois which contained contact information, pedigree information and family information for Canavan families worldwide and claimed that MCH and Matalon converted the names on the register and the genetic information by utilizing them for the hospitals' exclusive economic benefit. The court, however, dismissed the conversion claim by denying any property interest regarding the body tissue and genetic information voluntarily given to defendants. The body tissue and genetic samples were donations to research without any contemporaneous expectations of return.[72]

In the last claim, the plaintiffs alleged that MCH misappropriated a trade secret i.e. the registry of people who had Canavan disease. However, they failed to make

---

[66] *Ibid.*
[67] *Ibid.*
[68] *Greenberg, supra* note 59.
[69] *Ibid.*
[70] *Ibid.*
[71] *Ibid.*
[72] *Ibid.*

allegations regarding the facts that MCH knew or should have known that the Canavan Registry was a confidential trade secret guarded by Plaintiffs, and furthermore, that Matalon had acquired it through improper means. This deficiency had ruled out any possibility of misappropriation of trade secrets. Accordingly, the court dismissed the claim by mentioning that plaintiffs have failed to state a claim regarding misappropriation of trade secret as they have not sufficiently alleged the requisite elements to convert the Registry into an actionable trade secret.[73]

The decision in the Greenberg case elaborates the principle, set forth by the Supreme Court of California in its 1990 decision, *Moore versus Regents of the University of California*, that un-consented research uses of human tissue (in that case, the plaintiff's splenetic tissue) to develop commercial products from which contributors of the tissue will not benefit can give rise to a cause of action against the developer of the commercial products. Although, both the cases differ significantly regarding the cause of action as in the Moore's case the court recognised breach of fiduciary duty and lack of informed consent as cause of action, as opposed to the unjust enrichment claim in Greenberg.[74]

Both the parties to the Greenberg i.e. the plaintiff and defendant reached a confidential settlement effective from 6 August 2003, that allows the researcher and hospital to retain the patent at issue but that requires them to license the use of the patent to other Canavan's disease researchers without charge. Moore and Greenberg, both the cases have potential ramifications with respect to the sufficiency of consent for secondary uses of tissue and data collected during the course of a research study.[75]

### 6.2.3 Washington University Versus Catalona

Repositories of biological samples and clinical data are gaining utmost importance in biomedical research due to their potential use. They are collected in different ways: in the course of clinical care, previously conducted research studies, tissue requisition campaigns and other sources. Due to variety of ways in which biological samples and data are collected, the determination of their ownership becomes very confusing. 'Who owns biological samples and data housed in these repositories—the individual from whom it is harvested, the physician or researcher who draws it, or the university or institution that provides facilities, funding or oversight for clinical care and research?'[76] A satisfactory answer to this question remains to be found. *Washington University versus Catalona*[77] reflects the said confusion regarding the ownership of human tissues and data.

---

[73] *Ibid.*

[74] Barnes and Heffernan, *supra* note 19 at 4.

[75] *Ibid.*

[76] "Ownership of Biological Sample and Clinical Data II: U.S. Supreme Court denies certiorari in the Catalona decision", available at www.mwe.com. (visited on May 30, 2012).

[77] *Washington University v Catalona* 437 F. Supp. 2d 985 (E.D. Mo. 2006), *aff'd*, 490 F.3d 667 (8th Cir 2007), cert. denied,128 S. Ct. 1122 (2008).

6.2 Ownership Rights of Research Subjects over Their Genetic ...

The case highlights the current legal uncertainty over who owns or has rights to information and tissues collected from patients or research subjects and maintained for possible future uses. In this case, the issue was whether patients who donate tissues to a researcher or university may lose all rights to those tissues or they retain some. In the early 1980s, Dr. Catalona, a faculty member of Washington University, started asking his patients to give consent to use their tissue he removed during their surgery for research. He obtained their consent and collected a large number of tissue samples. His research led to the development of the PSA (prostate-specific antigen) which is used to detect prostate cancer. In 2001, Dr. Catalona requested that a limited number of samples be sent to a biotech company to evaluate the effectiveness of a new test to identify prostate cancer.[78]

His intention behind the request was to use the research results in an upcoming academic meeting and publish the same in a medical journal. The university balked the request by saying that a research publication would have enriched the medical community and benefited cancer survivors and their families but selling the samples had none of these benefits. Annoyed by the university's interference with his research, Dr. Catalona decided to move his practice to Northwestern University's medical school in Chicago. He asked his patients to express their willingness to transfer their samples to Northwestern. Six thousand of his patients wrote that they wanted their samples to move with him for future research.[79]

In the beginning, when Dr. Catalona asked his patients if he could use their tissue, it was a common practice for university researchers to use the samples and take them along if they changed job locations. But as time passed, Dr. Catalona's employer, Washington University, began to realize that the tissue samples could earn money for the university. Realising the commercial potential of samples, the Washington University filed a lawsuit against Dr. Catalona, asking the court to declare it to be the owner of the research participants' samples. It also asked for an order preventing Catalona from interfering with the patients and their samples.[80]

Washington University alleged that Dr. Catalona improperly asserted ownership rights on a tissue repository maintained at the university containing specimens collected from patients and research subjects at the university.[81] The university contended in its claim that tissue samples from repository were collected from patients and research subjects with their informed consent. In this regard, patients undergoing surgery or otherwise receiving treatment signed a consent form stating that the patient 'consent[s] to the disposal, use or examination of any bones, organs, tissues, fluids or parts which it may be necessary to remove'.[82]

The university alleged in its complaint that Dr. Catalona improperly contacted prior research subjects and patients, without the university's permission, seeking their consent to transport their samples (and presumably, attached data) with him to

---

[78] "Patients lose law suit to reclaim their tissues", available at http://www.whoownsyourbody.org/catalona.html (visited on May 22, 2012).
[79] *Ibid.*
[80] *Ibid.*
[81] Barnes and Heffernan, *supra* note 19 at 4.
[82] *Id.* at 5.

a new position he accepted at Northwestern University's Feinberg School of Medicine. The university asked in its complaint for a declaration that any of the 'consents' of Dr. Catalona's former patients and research subjects obtained by Dr. Catalona to the transport of the materials are invalid. It also asserted that Dr. Catalona's contacting his former patients and research subjects to secure these consents was an illegal practice under HIPAA.[83] On the other hand, Dr. Catalona contended that the patients and subjects retained control over their samples and information even after they had donated them to the repository, and that, therefore, it is not within the university's power to prohibit the patients and subjects from authorizing Dr. Catalona to take the material from Washington University to Northwestern.[84]

The research participants argued that the informed consent documents they had signed agreeing to participate in research demonstrated that they retained ownership rights over their literal flesh and blood. The informed consent forms promised the men the right to withdraw from research at any time, and some even promised the right to have the sample destroyed upon request. They contended that if they had the right to destroy their tissue, clearly someone else did not own it. They asked to have the samples moved to Northwestern University so that Dr. Catalona could continue the agreed-upon research.[85] But Washington University asserted that it had the right to use the research participants' samples as it wished 'in its sole discretion'. In spite of the language in the informed consent documents allowing the participants to withdraw at any time and for any reason, the university argued that the patients had given a gift—a donation—to the university, and they could not take it back.[86]

The US District Court for the Eastern District of Missouri held that, on the basis of the facts presented, Washington University owns the biological samples and associated clinical data, and that research participants did not retain any ownership interest beyond the right to withdraw their samples from the repository. A trial judge ruled that the informed consent documents were 'inconsequential'. The decision had given free hand to the university to perform stigmatizing or ethically objectionable research on the participants' tissue, or sell it to a biotech company for profit. The university, therefore, successfully prevented Dr. Catalona from transferring the samples to Northwestern. Against this decision, Dr. Catalona made an appeal to the US Court of Appeals for the Eighth Circuit, and the Eighth Circuit upheld the District Court's ruling on 20 June 2007. On appeal, Washington University's attorney argued that research participants should have fewer rights than regular patients and that participant had no right to stop research on their bodies that they find objectionable.[87]

The Eighth Circuit held that individuals who make an informed and voluntary decision to contribute their biological materials to a particular research institution for the purpose of medical research do not retain an ownership interest allowing

---

[83] *Ibid.*

[84] *Ibid.*

[85] *Supra* note 78.

[86] *Ibid.*

[87] *Ibid.*

such individuals to direct or authorize the transfer of such materials to a third party. Therefore, the court ruled that the university held sole ownership of the biological samples, and Dr. Catalona could not transfer the samples to a third party, even with the donors' consent. The court also made it clear that under the specific facts of the case, the men who participated had given their tissue to the university as a gift and they could not get it back or have it sent to another researcher.[88]

The court did acknowledge that the research participants had a right to withdraw (i.e. by destroying) their samples from the database, but did not conclude that a right to transfer the samples can be implied from the right to withdraw. [who owns..] The court noted that all informed consent forms expressly stated that research participants could discontinue their participation by having their samples destroyed upon request, but they did not state that participants could request return or transfer of their samples. Furthermore, the court highlighted that some federal and state regulations prohibit the return of biological samples to research participants.[89]

The court relied on various factors in support of its decision; first, numerous material transfer agreements and research agreements that both the university and Dr. Catalona had signed acknowledged the university's sole ownership of the biological samples; second, the university's intellectual property policy, to which Dr. Catalona was subject when he collected the biological samples, stated that 'all intellectual property (including…tangible research property) shall be owned by the University if significant University resources were used or if it is created pursuant to a research project funded through corporate, federal or external sponsors administered by the University.' and third, the university provided the majority of the funding for the maintenance and operation of the repository, and the remainder of the funding came from public and private grants to the university.[90]

Dr. Catalona appealed the Eighth Circuit's decision to the Supreme Court of the USA, but the Supreme Court denied certiorari without comment on 22 January 2008. In essence, Dr. Catalona asked the Supreme Court to rule that the Common Rule prohibited the university from asking the research participants to waive their ownership rights, and, therefore, the research participants still owned the samples and could direct their future disposition.[91] The Eastern District of Missouri addressed the exculpatory language argument, finding that the Common Rule's exculpatory language prohibition was limited to waivers or releases of liability, not ownership. The Eighth Circuit had not considered the effect or validity of such waiver language, because it felt such analysis was unnecessary in the context of a gift. The research participants also filed a petition for writ of certiorari, asking the Supreme Court to consider the Common Rule issue from their perspective.[92] On 22 January 2008, the Supreme Court denied the petition for certiorari without

---

[88] *Ibid.*
[89] *Ibid.*
[90] *Supra* note 76.
[91] *Ibid.*
[92] *Ibid.*

a comment, leaving unresolved such key issues as the application of the Common Rule's prohibition against exculpatory clauses.[93]

The present case invokes a greater responsibility on research institutions to review their informed consent procedures and forms in order to properly characterise any transfer of biological samples or clinical data as a donative gift, and state to whom the materials are transferred in order to provide the institution with flexibility to use the biological samples and data; and make clear that individuals only retain the right to withdraw their biological samples from the research at any time, and the language that appears to waive the subject's rights should be carefully reviewed in light of the Common Rule's prohibition against exculpatory language.[94] The case also highlights the importance of laying the ground for any ownership claims through the use of express policy and agreement language. Institutions should carve out intellectual property policies and intellectual property agreement provisions cautiously so as it should expressly secure the institutions' ownership rights to biological samples and associated clinical data against clinicians and researchers.[95]

### 6.2.4 Havasupai Case

This case represents the vulnerability of tribal people regarding informed consent and their potential exploitation by researchers and research institutions. In the present case, the Havasupai Tribe possesses restricted gene pools, in which certain genetic diseases are at higher incidence as compared to the general urban population. The tribe have one of the highest incidences of type-2 diabetes anywhere in the world. In 1991, 55% of Havasupai women and 38% of Havasupai men were diabetic.[96]

In 1989, two tribe members approached an Arizona State University (ASU) faculty member, Therese Markow, asking for help to stem the tribe's high incidence of diabetes.[97] Members of the Havasupai Tribe had given DNA samples to university researchers in 1990, in the hope that they might provide genetic clues to the tribe's devastating rate of diabetes. Subsequently, they learned that their blood samples had been used to study many other things, including mental illness and theories of the tribe's geographical origins that contradict their traditional stories.[98]

Members of Havasupai Tribe alleged that researcher Therese Markow and a colleague originally presented their project to the tribal council as consisting of three elements: (1) 'diabetes education', (2) 'collecting and testing blood samples from

---

[93] *Ibid.*

[94] *Ibid.*

[95] *Ibid.*

[96] "Research without Patient Consent", available at http://www.whoownsyourbody.org/havasupai.html (visited on May 28, 2012).

[97] *Ibid.*

[98] Harmon (2010).

individual members to identify diabetics or persons susceptible to diabetes', and (3) 'genetic testing to identify an association between certain gene variants and diabetes among Havasupai people'.[99]

Havasupai Tribe's seven-member tribal council approved the researcher's project with the understanding that the study would only involve diabetes research. However, at the same time it seemed that some written consents had contemplated certain future uses of the data and tissue for general population research ('to study the causes of behavioral/medical disorders'), given the fact that most subjects had been consented only verbally, or with informed consent forms that over time had been misplaced.[100] Great confusion persisted in the factual records also presented by the experts hired by ASU as to whether the informed consent process for all subjects, and the pre-research consultations between the researchers and tribal leaders, included mention of the possibility that tissue from the subjects might be used in later research on topics not related to diabetes.[101]

In 2004, two related lawsuits were filed by the Havasupai Tribe against ASU, the ASU Institutional Review Board (IRB), the Arizona Board of Regents and three researchers. It was alleged in the said law suits that tribal members' blood samples and handprints, which were purportedly collected as part of a study of diabetes affecting the Havasupai Tribe, were mishandled and ultimately used in studies of schizophrenia, inbreeding and population migration without the tribal members' consent to such future uses.[102] The complaints filed by the Havasupai Tribe itself and 52 individual tribal members, listed six causes of action: (1) breach of fiduciary duty and lack of informed consent; (2) fraud and misrepresentation/fraudulent concealment; (3) intentional infliction of emotional distress; (4) conversion; (5) civil rights violations; and (6) negligence.[103]

Acting on a motion to dismiss the case, the Federal District Court dismissed three out of six causes of action i.e. lack of informed consent and breach of fiduciary duty claims as well as the fraud and misrepresentation/fraudulent concealment and conversion claims. The judge maintained that the members of the tribe had consented to the research, even if the researchers had made fraudulent representations about the true nature of the studies to induce them to give consent. He ruled out any fiduciary relationship between the parties, even if the members of the tribe trusted the researchers, because the researchers claimed to not accept the trust that had been placed in them.[104]

However, the judge allowed the plaintiffs to continue their case with their claims for the infliction of emotional distress, violation of civil rights and negligence. Later on, the plaintiffs voluntarily dismissed their federal civil rights claim, stripping the federal court of jurisdiction to hear the case, and the court sent the case back to the

---

[99] *Supra* note 96.

[100] Barnes and Heffernan, *supra* note 19 at 5.

[101] *Ibid.*

[102] *Ibid.*

[103] *Ibid.*

[104] *Supra* note 96.

Arizona State Court. Upon remand, two cases—one for the tribe itself, and one for individual members of the tribe—were consolidated into a single case against the Arizona Board of Regents (for ASU and U of A), Markow and members of Markow's research team. The claims from the plaintiffs' Second Amended Complaint have been pending in Maricopa County Superior Court. In the said court, the plaintiffs claimed four causes of action: (1) breach of confidential or fiduciary duty (including lack of informed consent); (2) fraud and misrepresentation/fraudulent concealment; (3) negligence, gross negligence, negligence per se; and (4) trespass (with respect to the blood samples and entry onto tribal lands).[105]

In the meantime, acknowledging a desire to 'remedy the wrong that was done', the university's Board of Regents on Tuesday, 20 April 2010 agreed to pay $ 700,000 to 41 of the tribe's members, return the blood samples and provide other forms of assistance to the impoverished Havasupai. Legal experts considered this settlement very significant as it clearly shows that the rights of research subjects can be violated if they remain ill-informed about how their genetic material might be used. The Havasupai case raised important question whether scientists and researchers had taken advantage of a vulnerable population.[106]

Several genetics experts and civil rights advocates feel that the case may also add fuel to a growing debate over researchers' responsibility to communicate the range of personal information that can be gleaned from DNA at a time when it is being collected on an ever-greater scale for research and routine medical care. Though the case suggests that due regard should be given to the concept of informed consent and individual autonomy of research subjects, there are many scientists who argue that the potential benefit from unencumbered biomedical research trumps the value of individual control. In this regard, it is worth mentioning that Havasupai people were not against the scientific research but against the way in which the things went wrong. 'I'm not against scientific research', said Carletta Tilousi, 39, a member of the Havasupai tribal council. 'I just want it to be done right. They used our blood for all these studies, people got degrees and grants, and they never asked our permission'.[107] The Havasupai settlement appears to be the first payment to individuals claiming that their DNA was misused. Therese Markow, the geneticist, defended her actions as ethical and suggested that those judging her otherwise, failed to understand the fundamental nature of genetic research, where progress often occurs from studies that do not appear to bear directly on a particular disease. However, her suggestion seems insufficient to justify the violation of informed consent and emotional distress given to Havasupai people. The situation demands transparency and clarity in the procedure.[108]

The aforementioned cases suggest, although under differing theories, a possible movement towards holding research institutions, investigators, and, possibly, sponsors responsible for failing to obtain consent from subjects prior to using the

---

[105] *Ibid.*
[106] Harmon, *supra* note 98.
[107] *Ibid.*
[108] *Ibid.*

subjects' information for purposes not outlined in the consent form.[109] The case law, however, denied any possibility of treating human body parts as a form of property. The courts considered that conferring property rights to research subjects over their extracted genetic material may interfere with advancements in medical science. This gives an impression that an individual whose body provided the sample has no property rights in them. He is also denied any intellectual property right either.[110] Courts recognized certain rights, which are generally grounded in the notions of the fiduciary duty that a doctor owes to a patient and are frequently centred on the doctor's obligation to obtain informed consent. The analysis of aforementioned cases concludes that humans do not have any particular right to their cells or to the information contained in their cells, outside of their relationship with health care providers.[111]

Professor Robin Feldman of Hastings College of the Law, however, maintains that under the property law concepts, individuals should have a continuing property interest in their own cells.[112] She explains that the person from whom the cells have been excised must have some rights with respect to such cells, and such rights cannot be completely explained through the fiduciary duty of the physicians.[113] Excised human cells are first of all a tangible property that must belong to an individual. The logical ownership of cells lies with a person who contributed his cells.[114] Concerns are being made that conferring property rights to individuals in their excised cells may endanger the basic concept of human dignity. This could encourage the commercialization of basic elements of human body and could also erode the reverence of human life that infuses a number of legal doctrines, particularly, those that forbid ownership of human beings in the form of slavery. Moreover, treating human cells as property could also give rise to a situation where those in financial need will be coerced into selling parts of their body, either by unscrupulous brokers or by the circumstances of their poverty, raising serious concerns regarding health.[115] Rejecting these contentions as superfluous, Professor Feldman contends that 'if our concern truly lies with health risks to those who contribute their cells to research, denying property rights is a remarkably overbroad solution to the problem'.[116] She considers it a paternalistic approach to deny individual's property rights in the cells of their body in order to protect them from coercion while allowing property rights in the same cells to research labs.[117]

Professor Feldman remarks: 'You have no property rights in the cells of your body when they are outside your body because we must protect you from economic

---

[109] Barnes and Heffernan, *supra* note 19 at 4.
[110] Feldman, *supra* note 20, at 1379.
[111] *Id.* at 1380.
[112] *Ibid.*
[113] *Id.* at 1383.
[114] *Ibid.*
[115] *Id.* at 1384.
[116] *Ibid*
[117] *Id.* at 1385

exploitation, but we are perfectly comfortable letting biotechnology companies and research labs profit from the transfer of such cells'.[118] She argues that if we consider that no one should have property rights in human cells, the material transfer agreements highlight the property-like nature of those cells. This justifies the fact that someone must have legal dominion over excised human cells as tangible items.[119]

Although, research protocols have adapted to the need for informed consent, however, there is a dire need of creating indemnification agreements for cells donated. In the utmost enthusiasm of scientific advancements, we should not forget the necessity of thinking through the interests of the people, whose cells provide the raw materials. We must ensure that those raw materials are properly obtained. Now, when we have enormous experience regarding this type of scientific research, courts in the appropriate jurisdiction should revisit these issues in tune with the recent scientific and technological development.[120] Patent rights should be limited so that it will ensure a nice balance between the interests of society and access for scientific research and innovation and provide benefit to the individual.[121] Prof. Feldman concludes:

> A careful and more considered application of both property and intellectual property doctrines should provide more appropriate respect for the disparate interests in a way that is consistent with the doctrinal and theoretical roots in both areas. Perhaps in the end, it is our body, anyway.[122]

## 6.3 Ownership of Human Genetic Resources—from a Global Genetic Commons to National Property

Ownership of human genetic resources have become one of the contentious issues relating to human genetic material due to the increased extension of patent rights over human genetic resources. Ownership issue relating to genetic resources first started at the international level in 1983 when approximately 100 nations adopted the International Undertaking on Plant Genetic Resources (hereafter Undertaking), a nonbinding agreement negotiated under the auspices of the United Nations Food

---

[118] *Ibid.*

[119] *Ibid*; For example, in India, the Biodiversity Act 2002 recognizes sovereign rights over non-human genetic resources. *See* the Preamble and Sec. 2(c) of the Biodiversity Act 2002; Further, Department of Biotechnology, Ministry of Science and Technology constituted National Bioethics Committee in 1999 that has formulated Ethical Policies on the Human Genome, Genetic Research & Services. These policies emphasize that "International Law allows for the identification of ownership of sovereign rights over human genetic material (like any other biodiversity plants, animals and microbes) which shall be implemented." *See* Department of Biotechnology, Ministry of Science and Technology, *Ethical Policies on the Human Genome, Genetic Research & Services 1999*, available at http://dbtindia.nic.in/uniquepage.asp?id_pk=113 (last visited on May22, 2012).

[120] *Ibid.*

[121] *Id.* at 1402.

[122] *Ibid.*

and Agriculture Organization.[123] While considering genetic material as a part of global commons, the Undertaking states that it 'is based on the universally accepted principle that plant genetic resources are a heritage of mankind and consequently should be available without restriction'.[124]

### 6.3.1 Genetic Resources: A Global Genetic Commons

Although application of the Undertaking was restricted to plant genetic resources, however, all genetic material had traditionally been viewed as a part of global commons. In principle, as a part of a global genetic commons genetic resources were freely available to all.[125] They were freely accessible goods like information in public domain.[126] Like the living resources of the high seas they were not subject to sovereignty of any state.[127] In practice, while researchers were allowed under global genetic commons to collect samples of genetic material freely, however, there were two exceptions:

The open system did not grant researchers and scientists the right to trespass on private or state property to obtain genetic samples. Researchers had to obtain any consent normally required before entering such property. Also, researchers would pay collectors of such material for this kind of service. But they were not obligated to obtain national government approval for sampling activities or to compensate the country where the material was found.[128]

Genetic material, therefore, was considered as a global commons and no single user and country could claim an exclusive right over it.[129]

### 6.3.2 Global Genetic Commons Tradition Eroded with the Expansion of Intellectual Property Rights over Genetic Material

Global genetic commons tradition was eroded by the increased extension of patents to living organisms and later to genetic material. Patenting of isolated and purified genes and genetic sequences has become a norm in most developed countries.[130] The global genetic commons tradition was also adversely affected by the increased

---

[123] Safrin (2004).
[124] *Id.* at 644.International Undertaking on Plant Genetic Resources for Food and Agriculture, Nov. 23, 1983, Art. 1., available at http:www.fao.org/ag/cgrfa/ia.htm. (visited on May22, 2012).
[125] *Ibid.*
[126] *Ibid.*
[127] *Id.* at 644–645.
[128] *Id.* at 645.
[129] *Ibid.*
[130] *Ibid.*

assertion and expansion of other forms of intellectual property rights over plants, known as plant breeders' rights.[131] Furthermore, in 1989, certain developed countries exerted pressure for the addition of Annex 1[132] to the Undertaking to make clear that the Undertaking's common heritage of mankind concept did not affect the rights of plant breeders to exclude others from using their new and distinct varieties under the International Convention for the Protection of New Varieties of Plants (UPOV Convention).[133] As a result, in 1991, the UPOV Convention was revised to expand these breeders' rights by curtailing exceptions that had been allowed for the free replanting, exchange and use for breeding purposes of protected varieties and their propagating material.[134] These exceptions have the indicia of open system as they had allowed protected varieties to be used for a range of purposes without the original breeders' authorization. By the early 1990s, a variety of intellectual property rights were extended to biological goods and developing countries were pressurised to extend intellectual property protection to such goods.[135]

### 6.3.3 The Reaction of Developing Countries Against the Extension of IPR over Genetic Material

In response to the persistent extension of patent coverage to different life forms and realising the commercial potential of genetic material, developing countries committed themselves to assert sovereign rights to such material.[136] Here, the concern was that individuals and companies from gene-poor developed countries sought genetic material free of charge from gene-rich developing countries and then patent them and sold the same back to the country where the genetic material originated.[137] This prompted developing countries to take steps to enclose raw genetic material.[138] This happened in 1991 when they pressed for the adoption of Annex III to the Undertaking. The Annex made it clear that the Undertaking's heritage of mankind concept was 'subject to sovereignty of the states over their plant genetic resources' and that 'nations have sovereign rights over their plant genetic resources'.[139] This

---

[131] *Id.* at 646.

[132] FAO Res.4/89 available at www.upov.int. (Member states are expected to grant and protect these rights at the national level.) Sabrina, Safrin supra note 123 at 646.

[133] Safrin, *supra* note 123 at 646; International Convention for the Protection of New Variety of Plants Dec 2, 1961, as revised Oct. 23, 1978, Arts. 5(1), 6(1) 33 UST 2703 available at http://www.upov.int [Member states are expected to grant and protect these rights at the national level.].

[134] UPOV Convention as revised Mar. 19, 1991, Arts. 14(1), 14(5), 15(1)(iii), 15(2), S. TREATY DCC No. 104-17 (1995), available at http:www.upov.int (visited on May25, 2012).

[135] Safrin, *supra* note 123 at 646.

[136] *Ibid.*

[137] *Id.* at 647.

[138] *Ibid.*

[139] FAO Res. 3/41, available at http://www.fao.org/ag/cgrfa/ia.html.

assertion of sovereign rights over raw genetic material has eroded the concept of global genetic commons paradigm.[140]

The adoption of Convention on Biological Diversity (CBD) at the Earth Summit in 1992 had strengthened the sovereign ownership over genetic resources. The CBD begins its discussion of genetic resources by proclaiming the sovereignty of nations over them.[141] Article 15(1) of the convention states: 'Recognizing the sovereign rights of States over their natural resources, the authority to determine access to genetic resources rests with the national governments and is subject to national legislation'.[142] The convention broadly defines 'genetic resources' as 'any material of plant, animal, microbial or other origin containing functional units of heredity' of 'actual or potential value'.[143] After acknowledging sovereign rights over genetic resources, the CBD requires parties to 'endeavour to create conditions to facilitate access' to such resources.[144] However, the trend of gene-rich developing countries has been to restrict and encumber access to raw genetic material within their borders. They particularly object developed countries' granting of patents to genes isolated from material that was taken from or originated in developing countries. They view such patenting as 'colonial-style taking or theft'.[145] In addition to this trend, the environmental community has also encouraged developing countries to pass legislation restricting access to genetic material.[146]

### 6.3.4 Andean Common System on Access to Genetic Resources (Common System)

The Common System under the Andean Pact shows the sovereign control over genetic material.[147] Under the Common System, 'ownership of raw genetic material and derivatives of such genetic resources, such as molecules, effectively vests with the nation-state, i.e., the national government (the state), rather than with the individual or indigenous community whose land or property houses the relevant genetic resource'.[148] The state under the Common System either expressly owns or exercises virtually complete control over such resources.[149] It also ostensibly owns the

---

[140] Safrin *supra* note 123 at 647.
[141] *Ibid.*
[142] Convention on Biological Diversity, June 5, 1992, Art. 15(1).
[143] *Id.* Art. 2.
[144] *Id.* Art. 15(2).
[145] Safrin *supra* note 123 at 648.
[146] *Ibid.*
[147] Andean Common System on Access to Genetic Resources, Decision 391 (July 2, 1996), available at http://www.comunidadandina.org/ingles/treaties/dec/d391e.htm. (hereinafter Common System).
[148] Sabrin, *supra* note 123 at 650. (See Art. 6 of the Common system which stipulates, "[t]he genetic resources and their byproducts, which originated in the Member Countries are goods belonging to or the heritage of the Nation…as stipulated in their respective legislation").
[149] *Ibid.*

genetic material of migratory species, such as migrating birds, in their territories. As regards to the ownership of genetic materials, state ownership applies to those genetic materials for which the states are countries of origin.[150] In order to access genetic material within an Andean Pact country, a researcher or institution must obtain the consent of the national government and enter into an agreement prescribing benefits to be received by that country for the accessed material.[151] Only the state is empowered to authorise such access. Furthermore, only the state is a party to the benefit-sharing agreement with the bio-prospector.[152] The application of the Common System is not restricted to foreigners seeking access to genetic resources in an Andean Pact country but to Andean Pact nationals and local institutions also.[153]

### 6.3.4.1 Indian Position on Non-human Genetic Resource: The Biodiversity Act 2002

India passed a comprehensive legislation titled Biological Diversity Act 2002, restricting access to genetic material in its territory.[154] The Act prohibits any foreign person[155] or foreign corporation from obtaining any biological resource occurring in India or knowledge associated thereto for research, survey or commercial utilization without the prior approval of India's National Biodiversity Authority.[156]

However, the regulation of bio-prospecting by Indian resident citizens and Indian corporations is left to sub-national (state) biodiversity boards and appears less restrictive.[157] Restricting the rights of foreigners further, the Act prohibits any foreign person or corporation from transferring the results of any research relating to biological resources occurring in, or obtained from, India without the prior approval of the National Biodiversity Authority.[158] Moreover, even if the National Biodiversity Authority approves the obtainment or transfer of any such resource or information, the applicant may not subsequently transfer that resource or information without the Authority's consent.[159]

---

[150] *Ibid.*
[151] *Ibid.* (See Common System; *supra* note 147, Art. 17).
[152] *Ibid.*
[153] *Id.* at 650–651.
[154] Biodiversity Act 2002, No. 18, Feb. 5, 2003.
[155] This refers to any person who is not a resident citizen of India (See *supra* note 154, Sec. 3(2)).
[156] See *supra* note 154, Sec. 3(1).
[157] *Id.* Section 22.
[158] *Id.* Section 4.
[159] *Id.* Section 20(1).

## 6.4 Human Genetic Material

Since human genome holds greater potential for drug development as compared to the genes of plants and animals. Researchers seeking genetic links to diseases and drugs to cure them are aiming at access to genetic resources. The contemporary laws and draft laws that restrict access to genetic material to obtain remuneration for the nation exclude human genetic material from their ambit. This has increased the uneasiness over the use of human genetic resources for scientific or commercial purposes.[160] The situation becomes more complicated when human genetic material comes from developing countries. In such a situation, the issue of fair and equitable sharing of benefits between developed and developing countries becomes vital.[161]

Despite a decision of the parties to the CBD to exclude human genetic material from the convention's treatment of natural resources, a trend towards viewing human genetic material as a natural or national resource is apparent.[162] Although, the current legal regime differentiates between human and non-human genetic resources, however, this distinction may not hold over time. It is expected that the current rules and approaches regarding the ownership and control of non-human genetic resources may apply to the non-human genetic resources. Some see CBD as a model for future legal regime governing the ownership, access and control of human genetic resources.[163]

Attempts have been made to extend sovereign control on human genetic material e.g. China has made such attempt.[164] In the justification of extending the sovereign or national control to the human genetic material, some argue that if in *Moore versus Regents of the University of California*[165], the University of California, a state institution, can own an individual's cell line, cannot the national government of China or another country own the cell lines of their populations? National governments, which have successfully extended its control to non-human genetic materials, may jump to human genetic material with the attendant risks to human liberty and autonomy.[166]

### 6.4.1 International Law's Mishandling of Genetic Material and Its Implications

Developing countries are reacting very sharply against the patent systems by enforcing their access restricting regime e.g. India law prohibits any person from

---

[160] Sabrin, *supra* note 123 at 661.

[161] *Ibid.*

[162] *Id.* at 662.

[163] *Id.* at 662–663.

[164] *Id.* at 662, (Reports on human genetic research in China trumpet the importance of ethnic diversity as a national resource, describing the distinct characteristics of China's numerous ethnic groups as a "goldmine" for population geneticists.).

[165] *Moore, supra* note 28.

[166] Sabrin, *supra* note 123 at 663.

applying for any intellectual property right anywhere to inventions based on research or information on a biological resource obtained from India without the prior approval of the National Biodiversity Authority.[167] The authority has the discretion to impose a benefit-sharing fee and/or royalty as a condition of approval.[168] In Brazil, applicants for patents and other forms of intellectual property protection must 'specify the origin of the genetic material and the associated traditional knowledge', presumably in their applications for such protections.[169]

In order to recognise their access-restricting regimes at the international level, developing countries are persistently demanding that developed countries require patent applicants to disclose the country of origin of any genetic material used to develop the item sought to be protected. Furthermore, they are also demanding that developed countries either refuse to enforce or refrain from granting patents to innovations, such as synthesized genes and bioengineered goods that utilize material that come from developing countries, unless it has been obtained in compliance with the country of origin's access-restricting laws.[170]

### 6.4.2 Sovereign Control versus Open System

The open system relating to genetic material, which was prevalent before the CBD, carried numerous advantages.[171] Under the open system, samples of biological resources containing genetic material, such as seeds, soil, leaves and animals, were freely exchanged both within and between nations. This international sharing had 'facilitated the conservation and improvement of genetic material, as well as fostered international scientific collaboration'.[172] Genetic resources are such type of good that more we share it, more we preserve it.[173] The work of the Consultative

---

[167] *Id.* at 665, Biological Diversity Act, *supra* note 154, Sec. 6(1).

[168] Biological Diversity Act, *supra* note 154, Sec. 6(2).

[169] Brazil, Provisional Measure No. 2.186-16 (Aug. 23, 2001), available at http://www.grain.org/brl/brazil-tk-2001-en.cfm [hereinafter Brazil Measure] Arts. 26, 30, 31.

[170] Sabrin, *supra* note 123 at 666. [For example, in connection with the 2003 meeting of the TRIPS Council, Bolivia, Brazil, Cuba, the Dominican Republic, Ecuador, India, Peru, Thailand, and Venezuela proposed amending the TRIPS Agreement to require, as a condition of patent acquisition, (1) the disclosure of the source and country of origin of genetic resources used in the invention, (2)evidence that the country of origin had consented to its extraction and use, and (3) "evidence of fair and equitable benefit sharing under the relevant national regime." They further proposed that failure by an applicant to provide this information should render the patent unenforceable." They argued that these amendments were "imperative to implement the TRIPS Agreement and the CBD in a mutually supportive and complementary. The African Group, which consists of all African nations that belong to the WTO, proposed a similar amendment to the TRIPS Agreement, arguing that the Agreement "has not provided adequate and equitable means to prevent patents mainly in developed Members that have amounted to and resulted in the misappropriation of genetic resources… from developing members."]

[171] *Id.* at 670.

[172] *Ibid.*

[173] *Ibid.*

## 6.4 Human Genetic Material

Group on International Agricultural Research (CGIAR) presents an excellent example of the international conservation and collaborative activity that flourished under the open system. It has been a great challenge to preserve the CGIAR system.[174]

Arguments are being made that 'nations and international work addressing access to genetic material should begin with the normative assumption of an open system and then justify sovereign ownership or enclosure'.[175]

Before the adoption of CBD, there had been a great debate among scholars as to who should own raw genetic material. Some argued in favour of vesting the ownership with sovereigns on the basis that the sovereigns were in the best position to negotiate lucrative deals for raw genetic material. They argued further that if sovereigns owned genetic material, they would have an incentive to conserve it. There were others, who contended that indigenous communities should own genetic material, as they were the ones who had traditionally conserved it. Still others were in the favour of vesting ownership rights in individuals.[176] Sabrina Safrin suggests that 'the answer to the question of ownership of raw genetic material should have been: whoever owns the res that contains the genetic material'.[177] It follows from this that 'the owner will differ in different situations'.[178] For example, sovereign enclosure would be justified where the sovereign owns the tangible property that houses the genetic material.[179] Sovereign enclosure would be more justified where the country is home to unique or rare genetic material.[180] 'As with all genetic material, while the likelihood is slim that unique or rare material would yield a commercially lucrative product, at least any value derived from it would not be further eroded by the presence of multiple suppliers or conflicting claims and contests between nations'.[181] Safrin suggests an approach, which involves 'a change in focus by national governments and the international community from one that stresses the obtainment of remuneration by developing countries for genetic material in its raw state to one that stresses opportunities for developing countries to add value to such material'.[182]

### 6.4.3 Optimum Way Outs

Among the systems discussed above, patent system and sovereign system both have overreached in permitting or asserting ownership rights to genetic material.

---

[174] *Id.* at 671.

[175] Id. at 680 (Such a change in international assumption would not necessarily require an amendment to the CBD because the Convention allows countries to refrain from asserting sovereign rights. See CBD, *supra* note 142, Art. 13(7).).

[176] *Id.* at 681–682.

[177] *Id.* at 682.

[178] *Ibid.*

[179] *Ibid.*

[180] *Ibid.*

[181] *Id.* at 682–683.

[182] *Id.* at 684.

The increased enclosure of genetic resources through both the systems obstructs the full enjoyment of citizens from developed and developing countries and leave the potential of genetic resources unrealised. It not only creates tensions between nations but threatens individuals and indigenous communities also. Furthermore, such enclosure also diminishes opportunities to conserve, expand and improve the global genetic pool.[183] Sabrina Safrin suggests the remedy to repair this situation. She suggests 'a bilateral framework of steps to be taken by the United States and by gene-rich developing countries that would reduce both private and sovereign enclosure of genetic material and create a more open system'.[184] In addition to this, she advocates for the earlier paradigm, an open system at the international level in which genetic resources were readily shared by all and exclusively owned by no one. She contends that while the said open system was far from perfect, however, was not without advantages.[185] She concludes: '[r]ather than continuing their companion hyper-ownership approaches, developed and developing countries, as well as those conducting international work on genetic resources issues, should strive to achieve a more open system.'[186]

## 6.5 Position of International Agreements on the Access of Human Genetic Resources and Benefit Sharing: Convention on Biological Diversity and TRIPS

### 6.5.1 Benefit Sharing

Benefit-sharing is seen as a viable solution to the accessibility problem. One of the three objectives of CBD is the fair and equitable sharing of benefits arising from the utilization of genetic resources (CBD, Art. 1).[187] Although the CBD highlights the need for benefit sharing with local and indigenous communities, it does not provide legal rights to local communities regarding their knowledge or genetic resources; legislation is left to national governments.[188]

#### 6.5.1.1 Andean Pact

An approach to benefit sharing that has been incorporated into national and regional legislation is exemplified in the Common System on Access to Genetic Resources (referred to as Common System), adopted by the Andean Pact. The latter is a cus-

---

[183] *Id.* at 625.
[184] *Ibid.*
[185] *Ibid.*
[186] *Ibid.*
[187] The Convention on Biological Diversity (CBD) Article 1.
[188] World Health Organization (2002).

toms union agreed at Cartagena, Colombia (therefore sometimes also known as the Cartagena Agreement) in 1969 among five countries: Bolivia, Chile, Colombia, Ecuador and Peru. Venezuela joined the union, but Chile left, in 1976.[189] In 1996, the Andean Pact countries adopted the Common System which establishes region-wide access and benefit sharing regimes. The Common System states that member countries have sovereign rights over the use and exploitation of their genetic resources, including derivatives of these resources, and the right to determine conditions of access. It also extends rights over an 'intangible component', which refers to any knowledge, innovation or practice associated with a biogenetic resource or its derivative.[190]

Under this system, researchers who access genetic resources in these countries are required to sign an access contract before the research is carried out. These contracts have to take into account the rights and interests of the suppliers of genetic resources, and their derivatives and related intangible components, and guarantee the equitable sharing of benefits derived from access to these resources.[191] The Andean Pact was brought in line with TRIPS with the adoption of Decision 486 on 1 December 2000, under which patenting of microorganisms was introduced.[192]

### 6.5.1.2 TRIPS

Despite this, TRIPS neither recognizes the contribution of traditional knowledge nor the need for benefit sharing. Proposals have been made that provisions for incorporating benefit sharing concerns are explicitly written into TRIPS to protect traditional knowledge and genetic resources. Further discussion needs to be developed at the WTO on this issue.[193]

### 6.5.1.3 United States National Bioethics Advisory Commission

Among other organizations, the United States National Bioethics Advisory Commission also recommends that prior agreements are forged between research sponsors and local representatives, so that countries can set their own priorities for research that is undertaken, based on the potential benefit that will be fed back to the community.[194]

---

[189] *Ibid.*
[190] *Ibid.*
[191] *Id.* at 143–44.
[192] *Id.* at 144.
[193] *Ibid.*
[194] *Ibid.*

### 6.5.1.4 The International Bioethics Committee of UNESCO

The International Bioethics Committee of UNESCO has supported the idea of benefit sharing and recognized the need for global benefit sharing agreements and the development of novel mechanisms of intellectual property management that uphold the public good.[195] As genomic research expands, the possibilities for work involving the resources of the developing countries will undoubtedly increase. Although some progress has been made towards the definition of benefit sharing, a great deal of more work is required, particularly to ensure that international research collaborations are based on benefit sharing programmes which do not have potential disadvantages for the health and economy of the developing countries.[196]

### 6.5.1.5 Convention on Biological Diversity

Although in 2002, the Bonn Guidelines on Access and Benefit Sharing were adopted under the CBD, however, their application to human health is limited because, as Sec. C/9 of the guidelines lay out, the scope of the guidelines is defined so as to explicitly exclude human genetic resources.[197]

### 6.5.1.6 Human Genome Organisation (HUGO)

In 2000, the HUGO Ethics Committee, in its statement on benefit sharing, considered benefit sharing specifically in the context of research involving the human genome. The statement recommends, among other things, 'that profit-making entities dedicate a percentage (e.g. 1–3 %) of their annual net profit to healthcare infrastructure and/or humanitarian efforts'.[198] In an editorial for *Science* published in the same year, members of the HUGO Ethics Committee offered three arguments to justify benefit sharing:

- 99.9 % of the human genome is common to all humans. This entails a responsibility, grounded on human solidarity, to share in the benefits of research based on this common good.
- There is a long legal history of viewing global resources, such as the sea, air and space, as common goods, to be equitably available to all humans, and protected for future generations. The human genome can equally be regarded as a common heritage.
- Vast differences in power and wealth between those conducting human genetic research and those providing genetic samples for such research, as well as the

---

[195] *Ibid.*
[196] *Ibid.*
[197] *Supra* note 55 at 32.
[198] *Ibid.*

potential for substantial profits, raise concerns of exploitation. The HUGO Ethics Committee sees benefit sharing as a way to address these concerns.[199]

### 6.5.1.7 Indian Position on Benefit Sharing Under Biological Diversity Act

The law, in turn, requires the National Biodiversity Authority to secure equitable benefit sharing for the use of 'accessed biological resources, their by-products,... and knowledge relating thereto'.[200] Where a biological resource or associated knowledge has been acquired from a specific individual, group or organization, the authority may, but need not, direct payment of moneys collected to such individuals, groups or organizations.[201]

The law does not require bio-prospectors to obtain the consent of affected individuals or groups before obtaining biological resources. However, the National Biodiversity Authority must consult with specially created local committees when making decisions 'relating to the use of biological resources' and associated knowledge.[202] Finally, the law prohibits any person, whether foreign or Indian, from applying for any intellectual property right in or outside India 'for any invention based on any research or information on a biological resource obtained from India' without the prior approval of the National Biodiversity Authority.[203]

## 6.6 Possible Solutions

There is a great need to provide compensation or property interests for those, whose samples are studied, and allow them at least a say in how a patented invention should be used in order to mitigate purely commercial motives that could hinder access. The 'sources' to genetic material must be seen more than just exploitable founts of 'raw' material, with legitimate interests in sharing in the benefit of research in which they have made a contribution, intellectually or materially.[204]

---

[199] *Id.* at 32–33.

[200] Biodiversity Act *supra* note 154, Sec. 21(1).

[201] *Id.* Section 21(3).

[202] *Id.* Sec. 41(2). India's law necessitates that every local municipality create a biodiversity management committee. *Id.* Sections 41(1), 2(h). These local committees may levy collection fees from any person accessing or collecting biological resources from areas within their territorial jurisdiction. *Id.* Section 41(3). The National biodiversity Authority must notify the public of any of its approvals. *Id.* Sections 19(4), 20(4).

[203] *Id.* Section 6(1).

[204] *Supra* note 55. at 33.

## 6.6.1 Companies Incorporating Benefit-sharing Clauses in Their Strategies

Various companies have incorporated benefit sharing into their strategies, most notably: the Canadian firm Newfound Genomics, which donates 1 % of its net profits to a charitable trust for the general population; Decode Genetics, which was granted the exclusive use of a centrally compiled population health database in Iceland in exchange for paying all related expenses incurred by the government for its building and maintenance, in addition to an approximately US $ 700,000 annual payment, and 6 % of its gross profit; and the University of Hawaii, whose researchers discovered a gene mutation responsible for a rare genetic disorder called pseudoxanthoma elasticum (PXE), and subsequently filed for and obtained the first patent in the USA with one of the patients' parents as co-inventor.[205] Patrick Terry, chairman of the advocacy group PXE International, opined: 'With the heavy stick of holding a patent on the gene, we can accelerate the research process, control royalty and licence fees and eliminate turf wars between researchers'[206] This reflects how patents can sometimes be used strategically to secure access to proprietary technologies.[207]

## 6.6.2 Certificates of Origin for (Non-human) Genetic Resources

Many suggest certificates of origin for (non-human) genetic resources as tools for global distributive justice. The certificate discloses the country of origin of genetic material and proof of prior informed consent. Countries, which are rich in biodiversity and genetic resources such as Brazil, the Dominican Republic and Peru, have argued that certificates of origin be legally enforced.[208] However, patent offices in industrialised countries have been fiercely opposing the suggestion to include certificates of origin as a legal requirement for patent applications.[209]

## 6.6.3 Whether an Amendment Required in Patent System or Outside of it

As regards to benefit sharing, it remains contentious whether resolving the question of benefit sharing requires making changes to the patent system, or whether there may be other methods of achieving greater equity through policies and approaches outside of patent law. Ultimately, most commentators arguing for benefit sharing

---

[205] *Ibid.*
[206] *Id.* at 33–34.
[207] *Id.* at 34.
[208] *Ibid.*
[209] *Ibid.*

are not arguing against the ownership of genetic knowledge, but against its misappropriation and subsequent failure to fairly compensate sources.[210] A review by the WHO's Human Genetics Programme Chronic Diseases and Health Promotion 2005 concludes:

Developing countries should consider establishing policies that encourage entities involved in commercial aspects of research to negotiate openly with foundations and disease-associated advocacy groups or local community leaders for equitable benefit sharing. These countries should contemplate creating standards to guide such negotiation, in addition to carefully assessing mechanisms for negotiating the distribution of benefits resulting from international human genetic research.[211]

## 6.7 Need for Modification in the Existing Patent Regime

Suggestions are being made as to that the modification to the existing patent regime would be beneficial to the promotion of equitable benefit sharing. Issues of morality, which have been given little importance so far in the application of patent law, may be integrated into patent legislation modifications so as to ensure patent law is guided by ethics. Until alternatives are formulated, patent system could fill the gap through *ordre* public and morality clauses. There is a need to discuss, analyse and evaluate all the policy options in the light of affordability and accessibility of developing countries, paying much heed to the key aspects of equitable access to genetic technologies for vulnerable populations.[212]

## References

Barnes Mark & Heffernan Kate Gallin (2004) The future uses' dilemma: secondary uses of data and materials by researchers and commercial research sponsors. Medical Research Law & Policy Report 1: 3

Jasper A. Bovenberg (2006) Property rights in blood gene & data: naturally yours? Martinus Nijhoff Publishers, Lieden, p. 1

Boyle James (2003) Enclosing the genome: what squabbles over genetic patents could teach us. In: F. Scott Kieff (ed.), Perspective on properties of the human genome project. Elsevier Academic Press, London (U.K.) p. 111

Richard A. Epstein (2003) Steady the course: property rights in genetic material. In: F. Scott Kieff (ed.), Perspective on properties of the human genome project. Elsevier Academic Press, London (U.K.) p. 156

Feldman Robin (2011) Whose body is it anyway? Human cells and the strange effects of property and intellectual property law. Stanford Law Review 63:1381

---

[210] *Ibid.*
[211] *Id.* at 35.
[212] *Id.* at 241.

Amy Harmon, Indian tribe wins fight to limit research of its DNA. New York Times, April 21, 2010. http://www.nytimes.com/2010/04/22/us/22dna.html. Accessed 22 May 2011

N. Narayanan, (2010) Patenting of human genetic material v. bioethics: revisiting the case of John Moore v. Regents of the University of California. http://www.ncbi.nlm.nih.gov/pubmed/20432879. Accessed 21 May 2012

"Ownership of biological sample and clinical data II: U.S. Supreme Court denies certiorari in the Catalona decision", available at www.mwe.com. Accessed 30 May 2012

Patients lose law suit to reclaim their tissues. http://www.whoownsyourbody.org/catalona.html. Accessed 22 May 2012

Research without patient consent available at http://www.whoownsyourbody.org/havasupai.html (visited on 28 May 2012)

Safrin Sabrina (2004) "Hyperownership in a time of biotechnological promise" American Journal of International Law 98: 645–46

World Health Organization (2002) Genomics and world health, Report of the Advisory Committee on Health Research 142. http://whqlibdoc.who.int/hq/2002/a74580.pdf. Accessed 1 June 2012

World Health Organization (2005) Genetics, genomics and the patenting of DNA—review of potential implications for health in developing countries 33 http://www.who.int/genpmics/Full-Report.pdf. Accessed 30 May 2012

# Chapter 7
# Conclusion and Suggestions

The conjunction of biotechnology and IPR has serious implications for law and society. Intellectual property laws, which were framed in industrial age, have proved to be inefficient in the present information age. Existing patent laws are confronted with new genetic inventions, which differ markedly from mechanical and chemical inventions that have been the traditional subject matters of patents. Modern biotechnology inventions, particularly genetic inventions, have become more valuable as an embodiment of information as compared to their physical attributes. The advent of bioinformatics and genetic databases demands a different patent approach, much in tune with the present information age. Although, it is not possible to create new IPR every time when a new technology emerges, however, fitting all sorts of inventions in a single set of law is also problematic.

Patents are essentially of territorial nature and grant of a patent is subject matter for national sovereignty. Given the territorial nature of patents, nations differ in the scope and coverage of patent protection. Even in nations having similar patent laws, courts differ significantly in their interpretation of those laws. Among the USA, European Union, Canada and India, the scope of patentable subject matter varies from one jurisdiction to another in legislative framework, patent practice and interpretation by courts. One of the common points between all the four jurisdictions is that they allow patents on biotechnology inventions including genes and gene fragments with different degrees of protection.

All the four countries have adopted almost similar approach regarding the patenting of microorganisms but differ significantly as to the patenting of macro-organisms (higher life forms). For example, genetically modified animals are patentable in the USA while in the European Union, a patent is granted on genetically modified animals only when the benefit to the mankind outweighs the suffering caused to animals. Moreover, in the European Union, patents on genetically modified animals were first restricted to transgenic rodents containing an additional cancer gene, rather than 'transgenic non-human mammals' and thereafter to transgenic mice. Canada has adopted a relatively restrictive approach regarding biotechnology patents as compared to the USA and Europe. Canada maintains a distinction between higher and lower life forms, permitting patents only to latter. In India, genetically modified multi-cellular organisms, including plants, animals and human beings, and their parts are not patentable.

The USA has adopted a relatively liberal patent approach among the four jurisdictions. The US patent law does not contain a list of exclusions, and judicially created exclusions such as laws of nature, physical phenomenon and abstract ideas have been interpreted liberally to expand the scope of patentable subject matter. As a result, fairly new subject matters such as software, business methods and methods of medical treatment have become patentable in the USA. Further, patenting of higher life forms, which faced great impediments in European Union and Canada, passed the patentability criterion without great hurdle. The *Diamond versus Chakrabarty*[1] decision of the US Supreme Court has widened the scope of patentable subject matter to a great extent by allowing patents on 'anything under the sun made by man', baring few exceptions. By making such interpretation, the court encompassed both foreseeable and unforeseeable subject matter including gene sequences. An imitation effect rippled from the USA to European Union and other jurisdictions, prompting a series of legislative measures to patent living forms. Moreover, Trade-Related Aspects of Intellectual Property Rights (TRIPS) internationalized biotechnological practices and enabled genetic engineering to yield important breakthroughs in the new millennium. Rebecca S. Eisenberg comments on the impact of *Chakrabarty* decision as:

The full consequences of the expansive approach to patent eligibility endorsed by the *Chakrabarty* majority continue to be felt far beyond the biotechnology industry…A quarter century ago it was unclear whether the subject matter boundaries of the patent system were expansive enough to embrace biotechnology and information technology. Today, it is not clear whether the patent system has any subject matter boundaries at all.[2]

Critics took the view that the *Chakrabarty* decision was a tactical victory for industry, where industry persuaded the court to secure patent protection.

Despite being a pioneer in the field of biotechnology and patent law, the USA has been struggling in providing an effective patent protection to human gene and gene fragment. The US patent approach has not been homogenous while dealing with the genetic inventions. Human gene became patentable in the USA after the *Chakrabarty* decision. Before granting patents on a gene, the United States Patent and Trademarks Office (USPTO) and courts in the USA have long recognised that isolated and purified chemical substances were patentable because they did not exist in nature. Applying the chemical analogy to biotechnology and following the same logic of isolated and purified natural substances, USPTO started issuing patents to isolated and purified human genes encoding protein drugs, diagnostic probes, gene replacement therapies etc. In 1991, in the case of *Amgen versus Chugai Pharmaceutical Co.*, the Court of Appeals for the Federal Circuit (CAFC) specifically held that 'a gene is a chemical molecule albeit a complex one.'

---

[1] 447 U.S. 303.

[2] Eisenberg (2006).

However, patenting of human genes and gene fragments faced a jolt when National Institute of Health (NIH) filed patent applications on Expressed Sequence Tags (ESTs) in 1991. EST controversy highlighted the character of genomic discoveries as information. After continued gene patent rush, the USPTO began receiving heavy criticism which forced it to strengthen the utility criterion by prescribing specific, substantial and credible utility. Despite a stringent utility criterion, the problem of gene patents has not been set to rest. The gene as information becomes a problematic candidate to pass the patentable subject matter criterion because information is excluded as abstract idea under the US Patent Act. The USPTO, initially resisted the extension of the patent protection to information technology, however, they gradually allowed the same. In *Street bank & Trust versus Signature Financial Group*,[3] the CAFC established the principle that an invention is patent eligible, if it produces a 'useful, concrete and tangible result.' The issue of patentable subject matter was specifically raised in Bilski, where the CAFC evolved new standards as machine or transformation (MOT) test, which prescribes that a claim contains patentable subject matter, if it is either tied to a particular machine or apparatus or it transforms a particular article into a different state or thing.

This decision has a great impact on diagnostic gene patents as a diagnostic gene patent which simply involves comparing a particular gene or a group of genes fails the MOT test. The reason for such failure is that 'the claim recites no physical or chemical alteration; nor does it refer anything changing reforming or otherwise becoming another thing. On appeal, the US Supreme Court held that MOT test is not the sole test for patent eligibility under Sec. 101. The court maintained that the test may be a useful and important clue or investigative tool; it is not the sole test for deciding whether an invention is a patent eligible process under Sec. 101.

The Breyer J. of the Supreme Court highlighted the ill effects of MOT in the present information age:

> The machine-or-transformation test may well provide a sufficient basis for evaluating processes similar to those in the Industrial Age—for example, inventions grounded in a physical or other tangible form. But there are reasons to doubt whether the test should be the sole criterion for determining the patentability of inventions in the Information Age. As numerous amicus briefs argue, the machine-or-transformation test would create uncertainty as to the patentability of software, advanced diagnostic medicine techniques, and inventions based on linear programming, data compression, and the manipulation of digital signals.[4]

In *Association of Molecular Pathology versus USPTO*,[5] the southern District of New York, Judge Sweet held that 'DNA represents the physical embodiment of biological information, distinct in its essential characteristics from any other chemical found in nature. It is concluded that DNA's existence in an 'isolated' form alters neither this fundamental quality as it exists in the body nor the information it encodes.' The court's stand that genes are distinct from chemicals was a radical departure

---

[3] *Street Bank & Trust Co. v. Signature Financial Group, Inc.*, 149 F.3d 1368 (Fed. Cir. 1998).
[4] *Bilski et al. V. Kappos*, No. 08–964 (2010) slip op. at 9
[5] *Association for Molecular Pathology et al. v United States Patent and Trademark Office et al.*, No. 09-CV 4515, (S.D.N.Y. March 29, 2010).

from precedent, as in *Amgen Inc. versus Chugai Pharm. Co.*,[6] the court held that '[a] gene is a chemical compound, albeit a complex one.' This challenged an established practice of USPTO of granting patents on isolated and purified genes. Regarding the method claims, the court applied that MOT test and concluded that none of the patent claims were tied to any particular machine nor did they bring about a tangible transformation of anything. It maintained that comparisons of DNA sequences involved in these patents are abstract mental process, therefore also unpatentable.

The CAFC reversed the District Court's decision that Myriad's composition claims to 'isolated' DNA molecules cover patent-ineligible products of nature under Sec. 101 since the molecules as claimed do not exist in nature. The court also reversed the District Court decision that Myriad's method claim to screening potential therapeutics via changes in cell growth rates is directed to a patent-ineligible scientific principle. However, the court affirmed the District Court's decision that Myriad's method claims directed to 'comparing' or 'analyzing' DNA sequences are patent ineligible since such claims include no transformative steps and cover only patent ineligible abstract mental steps.

In March, the Supreme Court rejected the Court of Appeal for Federal Circuit's (CAFC) holding in *Prometheus Labs Inc. versus Mayo Collaborative services*[7] and unanimously held that the personalized medicine dosing process invented by Prometheus is not eligible for patent protection because the process is effectively an un-patentable law of nature. It is expected that the Supreme Court will now vacate and remand the pending *Myriad* case with instructions to the Federal Circuit to reconsider its holding that isolated human DNA is patentable. Following *Mayo*, the court could logically find that the information in the DNA represents a law of nature, that the DNA itself is a natural phenomenon, that the isolation of the DNA simply employs an isolation process already well known and expected at the time of the invention, and ultimately that the isolated DNA is un-patentable because it effectively claims a law of nature or natural phenomenon. The lack of homogeneity in the patent approach of the USA, while dealing with human gene and genetic information reflects the need for a review of existing patent law in the light of present information age.

The European Union, Canada and India also allow patents on human gene and gene fragments. In the European Union, an element isolated from the human body or otherwise produced by means of a technical process, including the sequence or partial sequence of a gene, may constitute a patentable invention, even if the structure of that element is identical to that of a natural element.[8] In India, gene sequences and DNA sequences with disclosed function are considered patentable.

The USA and Canada do not contain *ordre* public and morality clause in their respective patent laws. In Canada, the issue of morality had been raised in *Harvard College* case, where the Supreme Court unanimously held that the Commissioner

---

[6] *Amgen Inc. v Chugai Pharma. Co.* 927 F.2d 1200 (Fed. Cir. 1991).

[7] *Mayo Collaborative Services v. Prometheus*, 628 F. 3d 1347, reversed.

[8] Rule 23 e (2) of the Implementing Regulations, supra note 413.

of Patents has no discretion to refuse a patent on the basis of public policy considerations independent of any express provision in the Patent Act.[9] Although there is no public order and morality clause in US patent law, however, the issue of morality has been raised in various cases including *Ex parte-Allen*,[10] by animal rights groups and environmentalists farmer groups etc. However, public order and morality have not been impediments in granting patents over life forms.

The human chimera episode shows that though US patent law does not contain morality and public order provisions as a ground for rejection of patents, however, it recognizes its utility with respect to few serious issues. This stand of the USPTO towards patenting of human organisms has been given backing of law by the Leahy-Smith America Invents Act 2011. Section 33(a) of the said Act reads: 'Notwithstanding any other provisions of law, no patent may issue a claim directing to or encompassing a human organism.'

The issue of morality and public order has been often raised in biotechnology patent cases. However, *ordre* public/morality exemption is very rarely used to exclude an invention from the scope of protection. In Europe, a patent is granted over genetically modified animals only when the benefit to the mankind outweighs the suffering caused to animals. However, the potential unknown danger to society or environment will not prevent patenting of an invention. Human and human related patents are considered immoral and, therefore, not patentable in Europe. Since European Patent laws have specific statutory provisions regarding morality and public order, therefore, there is higher probability of rejecting patents on ethical and moral grounds in Europe than in the USA which lack such statutory provision.

Similar to Europe, India also recognizes public order and morality exclusion from patentability. It is apparent from the Indian Patent Act, 1970 that India strongly prohibits patents for genes and gene-based inventions based on morality and public order. Further, deep-rooted moral, cultural and religious beliefs in India set a litmus test for gene related inventions in order to pass the patentability criteria. So morality and public order plays an important role in determining patentability of gene-based inventions in India. Indian Patent manual 2011 echoes the same concern by mentioning that, 'An invention, the primary or intended use of which is likely to violate the well accepted and settled social, cultural, legal norms of morality, e.g. a method for cloning of humans.'[11]

The international patent regime is also struggling to deal with the recent biotechnology inventions. The lack of definition of terms like invention, microorganisms, microbiological processes and essentially biological process is creating uncertainty regarding the subject matter of patents. It has serious implications for developing countries because it provides enough space for technology rich developing countries under TRIPS to accommodate in their patent system a fairly new subject matter without having proper understanding. TRIPS does not distinguish between technologies;

---

[9] *Harvard College v. Canada (Commissioner of Patents)*, [2002 SCC 76, 116–121 per Bastarache J., and 89–102 per Binnie J.].

[10] *Ex Parte Allen* 2 U.S.P.Q.2d,1425, 2 (1987).

[11] *MPPP,* Chapter 8 Para 03.05.02 80–81(*2011).*

however, the genetic technology is posing unique challenges before the patent system. Here, it becomes pertinent to recognise the unique nature of gene patenting. However, the World Trade Organization (WTO) has some compulsions also as it recognises every new technology and permits discrimination, it would lead to great constraints in the path of patenting. I suggest that some explanatory clause be added to the Text of TRIPS which recognises the unique character of human genetic inventions. The flexibility provided under TRIPS, which allows states to interpret the term invention, is a right choice as it is difficult to demarcate a thin line between discoveries.

Professor Eisenberg rightly commented on the feasibility of the existing patent system in coping with the new technologies as 'the patent system was created for "a bricks and mortar world" which has inherent and logical limitations when transposed into the seemingly unlimited expansion of patentable subject matter.'[12] Like national patent laws, there requires a fresh approach regarding new technologies such as biotechnology at the international level. Flexibilities in the form of *ordre public* and morality clause are good because both the terms are relative terms.

The term '*ordre* public', derived from French law, is not an easy term to translate into English, and, therefore, the original French term is used in TRIPS. It expresses concerns about matters threatening the social structures which tie a society together i.e. matters that threaten the structure of civil society as such. 'Morality' is 'the degree of conformity to moral principles (especially good)'. The concept of morality is a relative concept which depends upon the values prevailing in a society. These values differ from society to society and country to country and also change over time. The decision as to the patentability of a subject matter sometimes depends upon the moral judgement of that subject matter in that society. In the light of this discussion, the *ordre* public and morality clause seems justified.

Article 27.3(b) of the TRIPS Agreement prescribes that microorganisms and microbiological processes shall be patentable in all the member countries but it neither defines these terms nor set any parameter to determine their scope. Although it is difficult to define microorganisms but there must be some parameters to determine the scope of it. There is a great uncertainty as to the scope of the term microorganism because even in scientific practice, the said term is inherently flawed as scientific classification continually evolves. In the medical literature, there is still no unanimously accepted definition of microorganism, neither a certain boundary between microbiology and biology. Microorganisms have been defined broadly as unicellular structures with much reduced dimensions and not visible to the naked eye. Another definition of microorganism is found in the decision of the Technical Board of Appeal of the European Patent Office in the *Plant Genetic Systems* case: 'the term "microorganism" includes not only bacteria and yeasts, but also fungi, algae, protozoa and human, animal and plant cells, i.e. all generally unicellular organisms with dimensions beneath the limits of vision which can be propagated and manipulated in a laboratory. Plasmids and viruses are also considered to fall under

---

[12] Eisenberg (2000).

this definition'[13] The European Union has decided to discontinue the use of the term 'microorganism'; instead, it has decided to use the term 'biological material', which means any material that contains genetic information and is capable of replicating itself or being reproduced in a biological system.

There has been a political divide regarding the extension of the concept of microorganism to animal cells between developing and most of the developed countries. Developing countries are against such extension, while many scholars in the industrialised countries advocate for maintaining the wording of Art. 27.3(b) of the TRIPS Agreement as such (i.e. without a precise definition) because microbiology is a fast-moving field of science and to provide a fixed and immutable definition of microorganism and microbiological process would reduce discretional powers of States at the implementation stage.

The draft of Art. 27.3(b) of the TRIPS Agreement indicates a temporary compromise among the many competing interests in the protection of biotechnology, which is evidenced by the inclusion of an early revision date for these provisions (January 1999). This Article is the sole provision in the TRIPS Agreement subject to an early review due to controversial nature of special protection given to some specific inventions. The framers of the said Article anticipated a negotiated revision of the terms of Article 27.3(b) as the primary way of resolving this controversy. The member countries seek modification and clarification of controversial terms via the WTO's administrative committees and dispute settlement procedures. Members differ regarding the review of the said provision, while some members maintain that the review should be an examination of the extent to which the current provisions have been implemented, others advocate for a more substantive process that might encompass changing the text of the article.

There has been no consensus thus far in the TRIPS Council on the nature of this review. Developed countries insist that Art. 27.3 merely speaks for a review of implementation while developing countries maintain that 'the review should include the possibility of revising the text not so much to strengthen but rather to loosen Art. 27's requirements with respect to patentable subject matter, and in any event to make this provision of TRIPS more sensitive to the needs and interests of the developing world.'

Developing countries propose that Art. 27.3(b) should be revised to ensure that naturally occurring materials, including genes, are not patentable. It should also recognise the adequate protection of traditional knowledge of local and indigenous communities. Developing countries seek that the exception for plants and animals should be maintained in the TRIPS. They also want that they should have the flexibility to develop sui generis regimes on plant varieties, suited to the seed supply systems of the countries concerned.

Developing countries further demand compliance with obligations contained in the Convention on Biological Diversity (CBD), particularly, to share the benefits with the country of origin of any patented biological material. This concern was

---

[13] *Green Peace Ltd v. Plant Genetic System N.V.* (Case no. T 0356/93-334 dated 21-02-1995) Point 34.

recognised in the Fourth Ministerial Conference of the WTO in Doha, where, the Doha Ministerial Declaration instructed the TRIPS Council to examine, inter alia, 'the relationship between the TRIPS Agreement and the Convention on Biological Diversity, the protection of traditional knowledge and folklore, and other relevant new developments raised by members pursuant to Art. 71.1 of TRIPS [which authorizes the TRIPS Council to undertake reviews in the light of any new developments which might warrant modification or amendment of this agreement].'[14]

However, under the current political climate where developed and developing countries differ significantly, it is very unlikely that exceptions from patentability permitted by Art. 27.3(b) will soon be eliminated. Simultaneously, it is also not likely that in upcoming round of multilateral trade negotiations at least some of the foregoing demands of the developing world will be met.

International patent community has given utmost importance to the harmonisation of patent laws, creating uniform patent law on a global scale through the diversity of the existing systems. The most significant manifestation of this trend is the TRIPS Agreement, which necessitates that the patent laws of the signatory nations must be in conformity with a uniform framework of international standards. Harmonisation of patent laws in the post-TRIPS world continues to be a shibboleth in patent circles, and diversity a flaw to be remedied. There is no doubt that harmonisation of patent laws provides certain benefits; however, it also entails certain disadvantages as it would preclude inter-jurisdictional competition and experimentation in patent law, among other things. Here, 'the relevant policy question is to what extent inter-jurisdictional diversity and competition should be sacrificed to achieve global uniformity.' This question is significant for understanding proper limitations of the measures already taken towards global harmonisation of patent law, especially the TRIPS Agreement. If jurisdictional diversity holds some merit, it becomes pertinent that the provisions under TRIPS permitting diversity and flexibility should be interpreted broadly. Harmonisation of patent laws has been more successful regarding the procedural aspects as compared to substantive aspects. Substantive harmonisation seems to be a far cry given the conflicting stands of developed and developing countries.

Patenting of human gene and gene fragments has not been as problematic as the effects it exerts upon genetic research and innovation on the one hand and various stakeholders of the society on the other. The social and policy implications of gene patenting cannot be adequately addressed by making changes in the patent systems rather by formulating policies and legislations to regulate the patent practices such as patent licensing etc.

Human genetic research is increasingly commercialised because of the potential benefits it creates for companies and biotech industries. It is raising valid concerns

---

[14] Doha WTO Ministerial 2001: Ministerial Declaration, WT/MIN(01)/DEC/1, Nov. 20, 2001, adopted Nov. 14, 2001, para. 17 and 19 [hereinafter Doha Declaration] para. 19; *See* also para. 17 (stressing the importance of implementing and interpreting the TRIPS Agreement "in a manner supportive of public health, by promoting both access to existing medicines and research and development into new medicines," in connection with which the Doha Ministerial issued a separate, and more detailed declaration.

regarding the free flow and open sharing of knowledge. The said commercialisation leads to a classification between most profitable areas and not so profitable areas. This has an adverse impact on the academic genetic research as academic researchers are now focused more on most profitable areas of research. This increased focus on commercialisation often lacks flexibility in objectives. There is always a possibility that the commercialisation of genetic research may lead to unnecessary delays in the presentation and publication of research results because researchers may be reluctant to share their research until the filing of its provisional patent application. This situation compels researchers to work in a closed environment as opposed to an environment, which allows open sharing of knowledge.

Sheer breadth of gene patents is also problematic as it produces an adverse impact on follow-up research. Recent studies show that too much patents can potentially deter innovation. Multiple patents on a single gene or gene fragment, which can be used as a research tool for further research, discourage genetic innovation. This is due to the fact that the accessibility of these tools demands negotiation with each of the patent holders. This can raise the transaction cost, leading to underuse of genetic resources. A survey[15] conducted by John Walsh reveals that the patent landscape has become more complex with more patents per innovation (including patents on research tools). The survey also confirms that patenting of upstream (basic) discoveries has increased, potentially limiting access for follow on research. However, on a surprising note, no respondents in the survey reported that worthwhile projects being stopped because of issues of access to IP rights to research tools. This is because university and industrial researchers have adopted working solutions that allow their research to proceed. The working solutions include licensing, inventing around patents, the development and use of public databases and research tools, court challenges, and simply using the technology without a license (i.e. infringement). The prevalence of licensing in drug industry suggests that the problem of access to patented research tools or upstream discoveries can often be settled contractually. However, the limitation with the contractual licenses is that they cannot obligate third parties. Despite these working solutions, aggressive assertion of IP can still have an adverse impact on scientific research. The most glaring example of such aggressive assertion of IP is Myriad's patent practice relating to the BRCA1 and BRCA2 gene. The exclusivity provided by aggressive patent licensing strategies may not be in the public interest: it may discourage others from working in an area which would profit from a variety of approaches or solutions. The situation demands a continuing need for active defence of open science.

Research exemptions have been seen as a viable option to ensure the accessibility of research tools to researchers and innovators. However, the imprecision of language and scope creates constraints. In the USA, the effective elimination of the research exemption by the recent CAFC's *Madey versus Duke University* decision may undermine the informal exemption, which is important for open science.

In recent years, numerous research organisations, research communities and private firms are arguing for the accessibility of genetic research tools to researchers

---

[15] Walsh et al. (2003).

and inventors. In the USA, the NIH does not favour patents on research tools obtained with federal funds and, if patents are obtained, favours their wide availability by non-exclusive licensing. Many private firms are committed not to file patent applications on research tools, and apparently ignore the patents of others. Isolated human genes and receptors are published but not patented as a matter of policy in many universities. Recent studies show that there is no sufficient evidence which indicates that the patenting over genetic research tools have a potentially deleterious effect on upstream research.

It has been the common practice within the research community for researchers to freely use research tools for academic purposes, provided patents for such tools are owned by an academic institution or individual researchers belonging to such an institution. Even in cases where a certain research tool is patented by a private sector company, the risk that researchers may face litigation as a result of using the tool in academic research has actually been extremely low; as private sector companies see little merit in exercising their rights against the use of their patented research tools in the academic research. However, the interviews conducted by Koichi Sumikura from 2004 to 2005 revealed the anxiety among researchers by using research tools in academic research has risen to unprecedented level.[16] The exclusivity provided by aggressive patent licensing strategies may not be in the public interest: it may discourage others from working in an area which would profit from a variety of approaches or solutions.

There is no effective data that gene patents actually inhibit genetic research. However, recent studies suggest that patenting of diagnostic genes have a deleterious effect on patients as it prevents patients from taking second opinion and verification testing. In the case of diagnostic gene patents, licensing practices are more important than the patent issue. The ultimate purpose of patents is to promote social good. However, patents, sometimes do not promote social good as it conflicts with the accessibility of genetic innovation. Innovation means delivery of goods, which raises concern that in the case of diagnostic patents whether patents of diagnostic tests lead to genetic innovation.

Intellectual property protection to bioinformatics and genomic databases has remained a debatable issue. Since the nature of bioinformatics is collaborative, therefore aggressive application of IPR may stifle with the process of innovation and access to genetic information. However, since bioinformatics is an emerging field, some sort of incentive structure is necessary in the form of intellectual property protection. Intellectual property protection to bioinformatics remains problematic as bioinformatics represents a fairly new discipline which combines two distinct fields viz. biotechnology and information technology (computer technology). Here, it is difficult to suggest which form of protection would be the most suitable protection because all the protections viz. patents, copyright and trade secret have some inherent limitations. In the case of bioinformatics database, the limitation with the patent protection is that only the method, and not the content, could be protected.

---

[16] Sumikura (2009).

In this situation, the protection provided by patent law is mere token protection and does not ensure total exclusivity of the data compiled.

Copyright protection could be extended to compilation. Facts are not copyrightable but compilation of facts is, provided there is sufficient degree of originality in the compilation in terms of selection and arrangement of terms, in terms of indices employed etc. However, the sui generis protection provided by EU Database Directive protects against unauthorised extraction of the information or utilisation of the whole or a substantial part of the database. Some suggests that the best form of protection which can be accorded to the bioinformatics database would be a combination of the traditional rights bestowed copyright law on compilation coupled with the rights guaranteed under the EU Directive. Such a combination would protect compilation to the extent of its selection and arrangement and also the contents from extraction and re-utilisation. The traditional rights, such as right to reproduction, licensing and publication would still apply and would be vested on the person making the compilation. However, the said combination would also be problematic as it would create confusion regarding the time period of protection.

Software now constitutes patentable subject matter under the US law, if it produces a useful, concrete and tangible result. Bioinformatics software is therefore eligible for the same protection as the software can be used for the purpose of biological research to produce results which are tangible, concrete and useful. Under Indian law, there is no patent available for a computer program *per se*.[17] However, the term '*per se*' is open to interpretation and a computer program coupled with some hardware component may fall under the scope of patentable subject matter, provided the claim is cleverly constructed in such a manner that the patent appears to be for the hardware, but the protection is claimed for the software as well, as an integral component.[18] In the realm of copyright, the term literary work has been construed to include software and protection accorded to software under copyright has been extended to human identifiable language, source code and machine readable component and object code.[19] Under the Indian law, software is included in the definition of literary work.[20] Also, computer program has been defined to include both source code and object code.[21] Protection is extended to computer program as long as the work is an original expression of the idea of the person creating the program.[22]

In Europe, a general-purpose computer programmed for a special purpose is, however, not excluded from patentability as long as it produces a technical effect. Copyright protection for computer software is not the best alternative, as the protection is same as extended to a literary work, and, therefore, extends only to the

---

[17] Sec. 3(k) of the Patents Act, 2005.

[18] *Supra* note 2 at 50.

[19] *Ibid*.

[20] Section 2(o) of the Copyright Act, 1957which includes computer databases under the definition of literary works. Cited in *Supra* note 2 at 50.

[21] Section 2(ffc) of the Copyright Act, 1957.

[22] Supra note 2 at 50.

original expression of the idea. Therefore, it is eminently possible for a person to merely change some aspects of the object and source code to claim an independent copyright, as long as it does not become a substantial copy of the original.[23]

However, in the realm of bioinformatics software, where there is a definite desire to market the product, there is possibility that the trade secret may be disclosed by reverse engineering. The object code may be used to reach the source code, and once this is done, protection effectively collapses. As regards to the IP protection to bioinformatics, the situation is in influx and the things will be clearer in due course of time as the field matures.

The completion of Human Genome Project unleashed enormous amount of genomic information. The immense value of this genomic information for prospective research necessitates that it should be made common and accessible to all humanity and would not be truncated, severed and owned by few individuals as first claimants. Researchers must have access to information resulting from the Human Genome Project as well as from subsequent research initiatives. However, since genomic sequences are potential sources of profit for the biotechnology and pharmaceutical industries, many private companies seek to limit access to this information. Since 1970s, a concerted effort has been made to make human genome sequence information freely accessible to researchers around the globe, and projects such as the Human Genome Initiative (HGI) have been created with this express purpose in mind. From the very outset, Human Genome Project (HGP) has emphasized that data obtained from HGP-funded research must be publicly available. However, a subsidiary, but explicit, goal of those responsible for creating and funding the HGP is the creation of technology and economic benefit. As a result, many members of the scientific community worry about the impact that such databases will have on research ventures. In recent years, there has been a push by commercial interests in the past decade to create private databases to recoup and profit from the investments of the private sector. Such databases are thus an important tool by which biotechnology companies make an early profit from genetic research. Without the promise of revenue, some argue, genetics and other fields of biotechnology could lose an enormous amount of funding from the private sector, and thus slow down the development of practical applications of genetic research. On the other hand, the HGP and organizations such as the NIH, Wellcome Trust, and other research groups argue that without public access, scientific research and advancement could be severely stifled.

It is important to note that some biotech firms are actually in favour of public databases. Patent law and economic realities ensure that private databases will continue to flourish, and so long as genetics and related technologies hold the promise of products and profit, there will be those interested in marketing the results. However, what has yet to be determined is the degree to which private databases control genetic information. A recent survey revealed that a growing number of human geneticists are against the privatization of genetic information. Groups such as the SNP Consortium Ltd. are taking aggressive action to protect certain information

---

[23] *Ibid.*

from being privatized, and other companies have acknowledged the importance of keeping genetic databases open to the public. Such actions reveal the strong feelings of the research community in favour of publically accessible data and against the further privatization of genetics.

It can be expected that as more genetic information is held by private interests, the tension between private and public sector groups will increase. Without the involvement of private companies, important therapeutic products may not reach the market. At the same time, private companies may not be able to develop these therapeutic products without the initial basic research provided by public research organizations. Increased collaborative efforts, such as that of the Wellcome Trust, are needed between private and public research groups to ensure continued advancements in genetic science and medicine.

The Bermuda Principles, which were aimed to facilitate the rapid release of genomic data, adequately address the problem of free riding by other scientists; however, they fail to address the problem of free riding by commercial researchers. To prevent this from happening, the HapMap Project developed a licensing strategy in the form of the HapMap Click Wrap License. To register for access to the HapMap Genotype Database, scientists must indicate acceptance of the HapMap Click Wrap License Agreement. By clicking on the acceptance button, they are granted a non-exclusive license to access and conduct queries of the Genotype Database. The license also includes the right to 'copy, extract, distribute or otherwise use copies of the whole or any part of the Genotype Database's data, in any medium and for all purposes', including commercial purposes.

The prevailing thought in legal discourse over IP and data is that data per se should not be protected. The general presumption is that academic research data form part of the public domain and that their accessibility is governed by the sharing ethos within the scientific community. However, the recent developments call for a robust intellectual property right in collection of data.

A solution to this problem is seen in open biotechnology development, popularly known as Open Bio development and sometimes open source biotechnology. Open Bio is a recent phenomenon in the field of biotechnology. The Open Bio movement is a reaction to the proliferation of IPRs and to concerns that IPRs may restrict research and access to new innovations.

Intellectual property proponents argue that Open Bio is antagonistic to IPR. However, an Open Bio project could include a mechanism to allow the initial researchers to recuperate reasonable production costs invested in its realization, without impeding the open nature of the project. This reflects that open biotechnology is not necessarily antagonistic to IP and that it is possible to develop an Open Bio project that would make use of the patent system. A variety of licensing schemes with or without IP (e.g., patent pool, non-assertion covenants, public domain, protected commons agreement, contractual licenses) can be used as effective tools to support the open nature of the project.

Open licenses are at the heart of any open project. They are the legal tools used to guarantee that the project remains accessible for all users and customers.

Unlike in IT, where most software is protectable through copyright, products of biotech are usually protected through the patent system. Moreover, several biotech developments initially thought to be protectable through the patent system have been found not deserving of such reward in recent judicial decisions, forcing developers to rely on other weaker IPRs (e.g. copyrights, sui generis database rights), contractual law or commercial secrecy for protection. Accordingly, it is extremely difficult to develop simple license models to ensure the openness of a given project and even more challenging to develop model licenses that could be used for a variety of projects.

Many small projects (private or public) simply cannot afford the cost of patents and prefer to rely on commercial secrecy to protect their inventions. An increasing number of scientists have turned to contractual licenses (often referred to as access agreements) to ensure open or controlled access to the fruits of their research to members of the scientific community. Purely contractual licenses, although less expensive and easier to design than patent licenses, are not particularly efficient against use by third parties to the original contract.

Open biotechnology is still in its infancy. However, the dynamism of the open biotechnology movement can be seen not only in the increasing number of open projects but also in the growing support and interest of policy makers, nongovernmental organizations and research funders, which bodes very well for the future of open biotechnology.

The effect on innovation will depend on conflicting influences: an open innovation process may lower the cost of research tool innovation by eliminating the transaction costs of license negotiations, but the potential benefits of inventive activity may no longer include possible profits from licensing the innovation. The core public policy justification of patents is that they stimulate innovation and diffusion by raising the private return to research, development and commercialisation. If open development lowers this private return, growth may suffer. Nevertheless, Open Bio may provide an opportunity for developing countries to imitate, learn and innovate without violating their IP agreements. In such fields, Open Bio effectively lowers the cost of entry into biotech research.

The legal question as to what ownership rights patients and research subjects have in their biological materials and their medical data is itself exceedingly ambiguous. The available cases[24] on the topic suggests a possible movement towards holding research institutions, investigators and possibly sponsors responsible for failing to obtain consent from subjects prior to using the subjects' information for purposes not outlined in the consent form. The case law, however, denied any possibility of treating human body parts as a form of property. The courts considered that conferring property rights to research subjects over their extracted genetic material

---

[24] *Moore v. Regents of the University of California*, 51 Cal. 3d 120 (1990); *Greenberg v. Miami Children's Hospital*, 2003 WL 21246347 (S.D. Fla. May 29, 2003); 264 F. Supp. 2d 1064 (S.D. Fla. 2003); *Washington University v Catalona* 437 F. Supp. 2d 985 (E.D. Mo. 2006), *aff'd*, 490 F.3d 667 (8th Cir 2007), cert. denied,128 S. Ct. 1122 (2008)

may interfere with advancements in medical science. This gives an impression that an individual whose body provided the sample has no property rights in them. He is also denied any IPR either. Courts recognized certain rights, which are generally grounded in the notions of the fiduciary duty that a doctor owes to a patient and are frequently centred on the doctor's obligation to obtain informed consent. The analysis of the cases concludes that humans do not have any particular right to their cells or to the information contained in their cells, outside of their relationship with health care providers.

Ownership of human genetic resources remains one of the contentious issues relating to human genetic material due to the increased extension of patent rights over human genetic resources. The current practices of potential exploitation of human genetic material through IPR have generated increased opposition to vital population genetic studies and other work of direct medical benefit to many countries. Benefit-sharing is seen as a viable solution to these problems. One of the three objectives of the CBD is the fair and equitable sharing of benefits arising from the utilization of genetic resources (CBD, Article 1). However, CBD is only concerned with non-human genetic resources and does not include human genetic resources. Despite a decision of the parties to the CBD to exclude human genetic material from the convention's treatment of natural resources, a trend towards viewing human genetic material as a natural or national resource is apparent. For example, reports on human genetic research in China trumpet the importance of ethnic diversity as a national resource, describing the distinct characteristics of China's numerous ethnic groups as a 'goldmine' for population geneticists. TRIPS does not recognise the contribution of traditional knowledge, or the need for benefit sharing. There have been proposals that provisions for incorporating benefit sharing concerns are explicitly written into TRIPS to protect traditional knowledge and genetic resources.

While present legal regimes distinguish between human and non-human genetic resources, this distinction may not hold over time. The rules and approaches currently being laid down, both nationally and internationally, regarding national government ownership and control of non-human genetic resources so as to obtain financial and other benefits for the nation can probably be expected to apply to or materially affect future legal regimes governing access to human genetic material as well. Indeed, some are already suggesting that the CBD govern or the parties otherwise address access to human genetic material. Others, while not going that far, point to the CBD as a model. National governments, having gained comfort from their ownership or extensive control of non-human genetic resources to obtain benefits for the nation, can make the jump to human genetic material, as the patent system has done, with the attendant risks to human liberty and autonomy. An open system, in which genetic resources were readily shared by all and exclusively owned by no one, though far from perfect, holds some merit.

## 7.1 Suggestions

On the basis of the present research work, following suggestions are being made:

### 7.1.1 Chapter 2

1. Biotechnology inventions are posing new challenges before the existing patent laws. These laws are struggling to cope up with these challenges as new biotechnology inventions differ markedly from those prevalent in the industrial age. This requires a comprehensive review of existing patent laws to address the genetic inventions in tune with the information age. The lack of such approach may prevent some useful inventions from society. The nature of the biotechnology inventions warrants altogether, different approach.
2. Biotechnology patents should be reviewed keeping in mind the ultimate purpose of patents i.e. to promote social good. Patentable criteria should be applied more stringently so that patents would be avoided to undeserved patentees. The USA presents an example that a flexible approach towards new technologies such as biotechnology produces useful results and also gives space to biotechnological inventions to settle down. This flexibility should be maintained while dealing with biotechnology inventions.
3. The patent approach should always follow the socio-economic conditions of a particular country and should be defined in respective socio-economic conditions. The *ordre* public and morality clause should be invoked to deny harmful biotech inventions.
4. Isolated and purified genes should be excluded from the purview of patenting as they perform almost same function in or outside of the body. The process of isolation, which was once a skilful and laborious manual task, has recently become a routine. This necessitates that isolation process should not be considered as a task of human ingenuity.
5. While making a distinction between patentable and non-patentable subject matter, the degree of human intervention is considered. It is an open question as to what extent human intervention is required to make a subject matter patentable. Here it should be considered whether the inventor is contributing significantly or not.
6. Process patents, which establish an association between a gene and a disease, should not be made patentable as it comes under natural phenomenon.

### 7.1.2 Chapter 3

1. Diversity and uniformity both have some merits and demerits; however, it should be kept in mind that the basis of international patent regime is the patent

## 7.1 Suggestions

laws of member states. Therefore, while harmonising the patent laws to bring uniformity, importance should be given to the territorial nature of patents unless it unnecessarily impedes the global trade. It is an open question for policy makers to think about the extent of harmonisation and the degree of uniformity. However, it is suggested that absolute uniformity would destroy the basis of patent law and caution should be taken to keep the diversity alive.

2. Flexibility should be kept intact but certainty to nebulous term must be provided because the uncertainty has a negative impact upon the interests of developing countries.
3. The lack of definition of terms like invention, microorganisms, microbiological processes and essentially biological process is creating uncertainty regarding the subject matter of patents. It has serious implications for developing countries because it provides enough space for technology rich developing countries under TRIPS to accommodate in their patent system a fairly new subject matter without having proper understanding.
4. TRIPS does not distinguish between any technologies, however, the genetic technology poses altogether unique challenges before the patent system. Here, it becomes pertinent to recognise the unique nature of gene patenting. It is suggested that some explanatory clause be added to the Text of TRIPS which recognises the unique character of human genetic inventions.
5. The flexibility provided under TRIPS, which allows states to interpret the term invention, is useful as it is difficult to demarcate a thin line between discovery and invention because some useful discoveries are being made patentable in the present information age.
6. There is a great uncertainty as to the scope of the term microorganism because even in scientific practice, the said term is inherently flawed as scientific classification continually evolves. In the medical literature there is still no unanimously accepted definition of microorganism, neither a certain boundary between microbiology and biology. In such a situation, it is difficult to define microorganisms; however, keeping in view the ill effects of such uncertainty, there must be some parameters to determine the scope of it. An explanatory clause must be added to the TRIPS in this regard.
7. Article 27.3(b) is the sole provision in the TRIPS Agreement subject to an early review due to controversial nature of special protection given to some specific inventions. The framers of the said article anticipated a negotiated revision of the terms of Article 27.3(b) as the primary way of resolving this controversy. The time is ripe to review this Article in the light of present technological developments. While reviewing this article, the concern of developing countries should be taken into account by maintaining sui generis option in the text.
8. Given the complex nature of biotechnology inventions and genomic inventions, there is a great need for setting higher standards for novelty and inventive step in order to ensure competition without violating international minimum standards and preventing routine discoveries from being patented.
9. Harmonisation process should have a cautious approach because of numerous factors including challenges posed by new technologies. The feasibility,

costs and benefits of a further harmonization should be adjudged from an economic as well as legal perspective. There is dearth of economic studies on the said topic. Further, there is no significant evidence in the favour of the thesis that patents produce development and the member states should adopt substantially the same standards of patent protection, irrespective of their level of development.
10. There is no doubt that harmonisation of patent laws provide certain benefits, however, it also entails certain disadvantages as it would preclude inter-jurisdictional competition and experimentation in patent law, among other things. Here, the relevant policy question is to what extent inter-jurisdictional diversity and competition should be sacrificed to achieve global uniformity. This question is significant for understanding proper limitations of the measures already taken towards global harmonisation of patent law, especially the TRIPS Agreement. If jurisdictional diversity holds some merit, it becomes pertinent that the provisions under TRIPS permitting diversity and flexibility should be interpreted broadly.

## 7.1.3 Chapter 4

1. The social and policy implications of gene patenting cannot be adequately addressed by making changes in the patent systems as patent law is not expected to provide solution to broad social and policy issues. Here, it becomes pertinent to formulate policies and legislations to regulate the patent practices such as patent licensing etc. in order to provide viable solutions to such issues. Furthermore, patent offices have a great responsibility to regulate the patent practices, promoting social good.
2. Patents on genetic research tools should be discouraged while applying patentable criteria stringently. If a patent is granted on a genetic research tool, a non-exclusive licensing over it should be encouraged. Non-exclusive licenses should be encouraged in diagnostic gene patents also. In this regard, it would be useful to apply the suggestion made by the Organisation for Economic Cooperation and Development (OECD) of a 'clearing house' to ease the obtaining of licenses for genetic inventions by commercial laboratories.
3. Policy makers should ensure an appropriate exemption for research intended for the public domain. Research exemption should extend to all uses of the patented inventions. Research exemption strategies should be revised in the light of present technological scenario, where, the line between academic (non-commercial) and commercial research is increasingly diminishing. Experimental use exemptions should be encouraged without making strict division between commercial and non-commercial research.
4. Despite various working solutions such as non-exclusive licensing, patent infringement, inventing around patents, the exclusivity provided by aggressive licensing practices retard innovation. In this regard, strategies should be made

to discourage aggressive licensing. Open science may be a better solution to this problem as it ensures lawful access to researchers without infringing patents.
5. Broad patent protection for genetic inventions may conflict with the innovation process as broad patent protection on a particular gene may limit opportunities for researchers, who want to carry on further research on that gene. Broad patent protection on genetic inventions, therefore, should be discouraged by patent offices by limiting the scope of patent claims.
6. Product patents on diagnostic genes should be discouraged because in a product patent the patent owner has exclusive rights to all subsequent uses of that gene. Patents on genetic research tools should be restricted to specific use ensuring the accessibility of such tools for further research. In this regard, Germany' purpose-bound patent protection can be a good guide.

## 7.1.4 Chapter 5

1. Bioinformatics and genomic inventions are new developments in the field of biotechnology and the nature of these inventions is significantly different from traditional biotechnology inventions. The emergence of bioinformatics and genomic databases has changed the face of biotechnology from lab-based technology to computer-based science. As a result, biological information and particularly genomic information has gained utmost importance. Since this information has been excluded from the purview of patenting as abstract idea, patents relating to bioinformatics and genomic inventions should be granted with utmost caution.
2. Intellectual property protection to bioinformatics and genomic databases remains problematic as patent, copyright and trade secret protections have some inherent limitations. Since this field is new, stringent application of any particular form of intellectual property protection would not be efficient. The situation is in influx and things would be clear in due course of time. As of now, further research is suggested to examine the nature of these inventions and viability of different forms of intellectual property protections.
3. The advent of bioinformatics and genomic discoveries necessitates a comprehensive review of existing intellectual property laws as to their relevance in the present informational age.
4. Open and collaborative efforts should be encouraged in the field of bioinformatics and genomic databases in order to ensure access to genetic information. Since the nature of bioinformatics is of collaborative nature, therefore aggressive application of IPR may stifle with the process of innovation and access of genetic information. However, bioinformatics as a new discipline also requires some incentive structure for its development in the form of IPR. The intellectual property approach to bioinformatics should be balanced in such a way that it should not only incentivise the inventor or creator but also ensure the open and collaborative nature of bioinformatics.

5. The OpenBio movement is a reaction to the proliferation of IPRs and to concerns that IPRs may restrict research and access to new innovations. That is why some biotechnologists see open biotechnology as antagonistic to intellectual property. However, open biotechnology projects could include a mechanism to allow the initial researchers to recuperate reasonable production costs invested in its realisation. It is, therefore, suggested that a mechanism should be developed, which is not antagonistic to IP and possibly would make use of the patent system. A variety of licensing schemes with or without IP should be used to support the open nature of a biotechnology project. Open licenses should be given more importance in open biotechnology projects so that they ensure accessibility of open biotechnology projects for all users and customers.
6. Open biotechnology provides an opportunity for developing countries to imitate, learn and innovate without violating their IP agreements. It may effectively lower the cost of entry into biotech research and foster local capabilities for innovation. However, open biotechnology development may reduce the profits from commercial R&D in the area of research tools. As a result, the potential benefits of inventive activity may no longer include possible profits from licensing the innovation. Here, it is suggested that developing countries should allow open biotechnology development after anticipating its potential and pitfalls in the light of their relative socio-economic condition and level of development.

## 7.1.5 Chapter 6

1. Ownership rights of research subject in their extracted genetic material must be recognised and if researcher or sponsor who is conducting the research gain any benefit, the equitable sharing of that benefit must also be recognised.
2. Research institutions and biotech companies involved in genetic research are indemnifying the research subjects through out of court settlements. However, the time is ripe to recognize the ownership rights of research subjects in the light of recent technological developments.
3. There is no doubt that medical science, especially genetic research cannot progress without the contributions from patients and research subjects; however, it is equally important that their interests should not be pitted against such progress. Attempts should be made to strike a balance between ownership rights of patients and research subjects and accessibility of scientific advances. A research subject, who supplies his genetic material for research is also making significant contribution and there is every reason to recognise his contribution.
4. Significant progress can be made by a more careful and considered application of current intellectual property doctrines. In this context, there requires a balanced intellectual property approach which pays utmost respect to the ownership rights of research subjects and the concepts of benefit sharing and informed consent in human genetic research.

5. Moreover, since, the existing international and national legal regimes dealing with non-human genetic resources such as CBD and Indian Biodiversity Act contain provisions relating to benefit-sharing arrangements for non-human genetic material; it becomes pertinent to draft laws which recognize the concept of benefit sharing in the context of human genetic material. Extending sovereign ownership to human genetic material, however, seems to be problematic as it may pose risk to human dignity, individual autonomy and human liberty. These attendant risks justify the distinction between human genetic material and other forms of genetic material and necessitates a specific approach regarding the ownership of human genetic material. CBD may be a model for drafting laws relating to human genetic material in few respects such as benefit-sharing and informed consent; however, the regulation of human genetic material demands specific legal approach.
6. *Havasupai* case reflects the vulnerability of research subjects and suggests that due regard should be given to the concept of informed consent and individual autonomy of research subjects. Many scientists maintain that the potential benefit from unencumbered biomedical research trumps the value of individual control. Here, it is noteworthy that Havasupai people were not against the scientific research but against the way in which the things went wrong. They needed transparency in clinical genetic research.
7. There is a great need to provide compensation or property interests for those, whose samples are studied, and allow them at least a say in how a patented invention should be used in order to mitigate purely commercial motives that could hinder access.
8. The sources to genetic material must be seen more than just exploitable founts of raw material, with legitimate interests in sharing in the benefit of research in which they have made a contribution, intellectually or materially.

## References

Doha WTO Ministerial 2001: Ministerial Declaration, WT/MIN(01)/DEC/1, Nov. 20, 2001, adopted Nov. 14, 2001

Eisenberg Rebecca S. (2000) Re-Examining the Role of Patents in Appropriating the Value of DNA Sequences. Emory Law Journal 49: 783

Eisenberg Rebecca (2006) The story of Diamond v. Chakrabarty: technological change and the subject matter boundaries of the patent system. In: Jane Ginsberg and Rochelle Cooper Dreyfuss (eds.) Intellectual Property Stories. Foundation Press, New York, p. 349

Sumikura Koichi (2009) Intellectual property rights policy for gene-related inventions-toward optimum balance between public and private ownership. In: David Castle (Ed.) The Role of Intellectual Property Rights in Biotechnology Innovation. Edward Elgar Publishing Limited, Cheltenham U.K./Massachusetts U.S.A., p. 88

Walsh John P, Arora Ashish, Cohen Wesley M. (2003) Working through the patent problem. Science 299: 14

# Index

**A**
Access and Cost of Patents, 165
Accessibility
 of abstract genomic data, 179, 180
 of genomic databases, 182, 183
 of research tools, 5, 153, 237
Access of Human Genetic Resources and
  Benefit Sharing, 222
Algorithms, 8, 172
 computer, 82
 mathematical, 41, 183
 patenting of, 174, 175
Analogies between Biotech and software, 191
Andean Common System on Access to
  Genetic Resources, 217, 218
Andean Pact, 217, 218, 222
Animals as patentable Subject matter, 32–34
Anticommons, 5, 141
Association for Molecular Pathology v.
  Myriad Genetics, 48, 49
Association for Molecular Pathology v.
  U.S.P.T.O., 44, 45

**B**
Bayh-Dole Act, 4, 24, 138, 149, 150
Benefit sharing, 9, 15, 222–226, 243
Bermuda principles, 180, 182, 241
Best mode, 35, 60, 64, 98, 106
Bilski Case, 42, 43, 46
Bioinformatics, 1, 8, 170, 173, 175, 176, 189
 databases, 8, 15, 171, 176, 192, 238, 247
 definition of, 170
 software, 171, 172, 177, 178, 185, 186,
  239, 240
Biotechnology, 1, 3, 10, 11, 15, 17, 20
 as a patentable subject matter, 23, 24
 in European Union, 67
 inventions, 2, 14
 patents, 22
BRCA1 and BRCA2 genes, 6, 49, 157, 160,
  161
Budapest treaty, 85, 106, 126

**C**
Canadian Institute of Health Research (CIHR),
  161
Canavan Disease Case, 203–206
Catalona case, 206
Certificates of origin for (non-human) genetic
  resources, 226
Chakrabarty Case, 31, 32
Chemical inventions, 20, 229
Commercialisation of Genetic Research, 237
Commercialisation of human genetic material,
  197
Community resource projects, 181–183
Compositions of matters, 25, 27
Compulsory licensing, 100, 162, 166
Consultative Group on International
  Agricultural Research (CGIAR),
  221
Convention on Biological diversity (CBD), 9,
  222, 243
Convention on the Unification of Certain
  Points of Substantive Law for
  Invention, 64
Copyright, 8, 173, 176, 184, 239, 240
 protection to software, 177, 178
Court of Appeal for Federal Circuit (CAFC),
  24, 46, 47

**D**
Databases, 188, 189, 237, 240
Deposition of biological material, 126

Diagnostic gene patents, 5, 43, 47, 63, 164, 231
Diagnostic genetic testing, 163
Diagnostic tests, 39, 42, 142, 153, 157, 163
Differentiation, 13, 127
Dimminaco Case- Paving the Way for Biotechnology Patents in India, 100–102
Direct to Consumer campaign bypass physicians, 164
Divergence in Biotechnology Patent Practices, 22, 23
DNA sequences, 21, 38, 45, 46, 58, 60, 61, 105, 106, 144, 147, 154, 165, 174, 232
Draft substantive patent law treaty (SPLT), 4, 129

**E**
Enablement, 60–64, 77, 106
Essential biological processes, 67
European Biotechnology Directive, 67, 115
European Database Rights Directives, 174
Exceptions under TRIPS, 116, 117
Expressed Sequence Tags (ESTs), 5, 150

**F**
Free and open source software (FOSS), 183, 189

**G**
Gene fragments, 2, 4, 5, 21, 39, 40, 141, 142, 229, 231, 232, 236
Gene patents, 2, 3, 12, 13, 22, 37–39, 63, 99, 156, 160, 237
General Agreement on Tariffs and Trades (GATT), 112
  negotiation, 123, 124
Gene sequencing, 2, 21, 38
Genetic information, 7, 8, 69, 79, 119, 155, 166, 179, 183, 205, 235, 238, 247
Genetic research, 4, 5, 9, 14, 15, 73, 137, 138, 142, 145, 153, 165, 198, 212, 236, 238, 240, 248
  importance of patents in, 139
Genetic Resources- A Global Genetic Commons, 215
Genomic databases, 8, 15, 60, 238, 247
  goals of, 179
Genomics, 1, 2, 22, 39, 60
Guidelines for the Examination in the European Patent Office, 2001 (EPO Gudelines), 66

**H**
HapMap project, 181, 182, 241
Harmonization, 4, 129, 246
  obstacles in, 132
Harvard College Case (Harvard Oncomouse), 82, 94, 232
Havasupai case, 210–214, 249
Human Chimera and Humanoid, 34
Human gene, 2, 12, 13, 20, 36, 37, 79, 160, 196
Human Genetic Patents, 20
Human Genetic Resources, 9, 15, 196, 214, 224, 243
Human Genome as Common Heritage, 198
Human Genome Organisation (HUGO), 224, 225
Human Genome Project, 2, 7, 39, 169, 179, 186, 197, 240

**I**
Impact of patenting of genetic research tools on innovation, 144, 145
Impact of patents with broad scope on genetic research, 140
Implications of patents relating to genetic research tools for society, 143, 144
Importance of Patents in Genetic Research, 139
Indian Patent Act 1970, as amended in 1999, 2002 and 2005, 100
Indian Position on Benefit Sharing under Biological Diversity Act, 225
Industrial applicability, 2, 74
Informed consent, 15, 134, 200, 202, 204, 207, 208, 210–214, 243
Innovation and open development, 191, 192
Intellectual property rights (IPR), 1, 2, 8, 9, 11, 12, 14, 112, 124, 196, 199, 213, 229, 243
Interfaces, 175
International Depository Authority (IDA), 106, 107, 126
International patent regime, 3, 14, 233, 244
International Undertaking on Plant Genetic Resources (Undertaking), 214
Invention, 2, 8, 14, 22, 24, 50–53, 64, 71, 80, 100, 102, 245
Inventive step, 2, 21, 67, 71, 73, 104, 105, 113, 245
Isolated and purified genes, 46, 47, 215, 232

# Index

## L
Legal implications, 4, 13
Life forms, 3, 20, 23, 25, 80, 216, 233
Limitations of a license, 183
Limitations on the Scope of Patentable Subject Matter, 28, 29

## M
Microbiological processes, 67, 99, 103, 115, 118, 119, 133, 108
Microorganisms, 3, 14, 18, 20, 27, 32, 82, 90, 103, 117, 234
Monsanto case, 94
Moore case, 199, 200, 202, 206
Morality, 3, 35, 77, 78, 81, 99, 107, 107
Multi-cellular organisms, 67, 68
Myriad Genetics Inc, 7, 46, 156, 158, 159
Myriad's Patents on BRCA1 and BRCA2 Genes, 154, 155

## N
Non-Human genetic resource, 15, 219, 243, 249
Non-obviousness, 2, 45, 50, 57–59, 97, 176
Novelty, 2, 20, 50–53, 70, 97, 132, 245

## O
Open and Collaborative Databases importance of, 188
Open Bio Good for Developing Countries, 192
Open bio movement, 183, 186, 241
Open biotechnology, 8, 15, 184, 186, 241, 242, 248
Open source biotechnology, 8, 241
Open standards, 188, 189
Ordre public and morality, 3, 22, 77–80, 117, 118, 232, 234, 244
Ownership of human genetic material, 8, 9, 195, 196, 249
Ownership rights of research subjects, 15, 248

## P
Paris Convention, 112
Patentability of biotechnology, 2, 65, 67, 123
  in European Union, 64
  in United States, 23
  inventions in Canada, 80, 81
  inventions in India, 99, 100
Patentability of genetic research tools, 143
Patentable subject matter, 2, 27, 32, 66–68, 81, 82
  plants as, 31, 36
Patent Cooperation Treaty (PCT), 126
Patent eligible subject matters, 2, 4, 40, 42, 48
Patent infringement, 25, 47, 93, 100, 149, 150, 152, 162, 246
Patenting of ESTs and Reach through Claims, 144
Patenting of genetic tests for diagnostic purposes, 153, 154
Patent law, 2, 3, 14, 17, 20, 85, 127, 130, 240
Patents rights, 79, 83, 91, 113, 140, 150, 152
Patent thickets, 5, 141, 145, 191
Phenotype data, 182
Pioneer Hi-Bred case, 86
Plant varieties, 89, 104, 120, 121, 235
Policy Implications of Myriad's Patents on BRCA1 and BRCA2, 163, 164, 165
Population genetics, 9
Product of Nature Doctrine, 23, 29, 198
Proteins, 1, 10, 38, 45
  endotoxin, 63
  therapeutic, 142

## Q
Quality of testing, 164

## R
Re Application of Abitibi Co, 82
Researchers, 6, 7, 9, 13, 14, 138, 146, 165, 190, 198, 211
  academic, 5, 9, 138
Research exemptions and their scope, 147, 148
Research subjects, 9, 196, 198, 207, 242
Research tools, 5, 14, 39, 142, 248
Royalty stacking, 5, 141

## S
Scope of the Patent Laws- Anything under the sun made by man, 28
Single nucleotide polymers (SNPs), 5
  consortium, 181, 190, 241
Social implications, 12, 162
Sovereign Control v. Open system, 220, 221
Status of biotechnology patent in India, 108
Statutory exception in developing countries, 152, 153
Sui generis system, 116, 120

## T
The Biodiversity Act 2002, 218
The European Patent Convention (EPC), 64, 123, 151, 173
The Genome Research and Accessibility Act (GRAA), 156

The International Bioethics Committee of UNESCO, 224
The Ontario Health Insurance Plan (OHIP), 162
Tightening of subject matter boundary, 42
Traces of a Unified System of Patents for European Union, 64, 65
Trade secrets protection, 178, 247
Tradition Knowledge Digital library (TKDL), 192
TRIPS agreement, 3, 4, 112, 115, 119, 235, 236, 246
TRIPS plus agreement, 128

**U**
UNESCO's Universal Declaration on the Human Genome and Human Rights, 1997, 197

Unicellular Organisms *See* Microorganisms, 67
United States National Bioethics Advisory Commission, 223
University-industry relationship, 4, 138
Utility, 2, 21, 39, 41, 45, 50, 54–57

**W**
Welcome trust, 190
World Intellectual Property Organization (WIPO), 4, 12, 129
World Trade Organisation (WTO), 112, 133, 134
Written description, 2, 35, 50, 60–62, 98, 126

The manufacturer's authorised representative in the EU is Springer Nature Customer Service Centre GmbH, Europaplatz 3, 69115 Heidelberg, Germany. If you have any concerns regarding our products, please contact ProductSafety@springernature.com

Printed and bound by CPI Group (UK) Ltd, Croydon, CR0 4YY

25/03/2026

02078174-0009